In-Vehicle Corpus and Signal Processing for Driver Behavior

NOTE TO READERS:

This book is supported by four websites:

1. DSP in Cars at SDSU (USA): Drive-Best
URL: http://dspincars.sdsu.edu/Drive-Best/

2. Driving Behavior Signal Processing at Nagoya University (Japan): Drive-Best
URL: http://www.sp.m.is.nagoya-u.ac.jp/NEDO/Drive-Best/

3. Center for Robust Speech Systems at UT Dallas (USA): Drive-Best
URL: http://www.utdallas.edu/research/utdrive/

4. Sabanci University in Istabul (Turkey): Drive-Best
URL: http://vpa.sabanciuniv.edu/Drive-Best/

Kazuya Takeda · John H.L. Hansen ·
Hakan Erdoğan · Hüseyin Abut
Editors

In-Vehicle Corpus and Signal Processing for Driver Behavior

Springer

Editors

Kazuya Takeda
Department of Media Science
Nagoya University
Nagoya, Japan
takeda@is.nagoya-u.ac.jp

John H.L. Hansen
Center for Robust Speech Systems (CRSS)
Department of Electrical Engineering
Erik Jonsson School of Engineering
and Computer Science
University of Texas at Dallas
Richardson, TX, USA
john.hansen@utdallas.edu

Hakan Erdoğan
Sabancı University
Istanbul, Turkey
haerdogan@sabanciuniv.edu

Hüseyin Abut
San Diego State University (Emeritus)
San Diego, CA, USA and
Sabancı University
Istanbul, Turkey
abut@anadolu.sdsu.edu

ISBN: 978-0-387-79581-2 e-ISBN: 978-0-387-79582-9
DOI 10.1007/978-0-387-79582-9

Library of Congress Control Number: 2008930874

Printed on acid-free paper

springer.com

Contents

1 **Improved Vehicle Safety and How Technology Will Get Us There, Hopefully** 1
Bruce Magladry and Deborah Bruce

2 **New Concepts on Safe Driver-Assistance Systems** 9
Sadayuki Tsugawa

3 **Real-World Data Collection with "UYANIK"** 23
Hüseyin Abut, Hakan Erdoğan, Aytül Erçil, Baran Çürüklü, Hakkı Can Koman, Fatih Taş, Ali Özgür Argunşah, Serhan Coşar, Batu Akan, Harun Karabalkan, Emrecan Çökelek, Rahmi Fıçıcı, Volkan Sezer, Serhan Danış, Mehmet Karaca, Mehmet Abbak, Mustafa Gökhan Uzunbaş, Kayhan Eritmen, Mümin Imamoğlu, and Çağatay Karabat

4 **On-Going Data Collection of Driving Behavior Signals** 45
Chiyomi Miyajima, Takashi Kusakawa, Takanori Nishino, Norihide Kitaoka, Katsunobu Itou, and Kazuya Takeda

5 **UTDrive: The Smart Vehicle Project** 55
Pongtep Angkititrakul, John H.L. Hansen, Sangjo Choi, Tyler Creek, Jeremy Hayes, Jeonghee Kim, Donggu Kwak, Levi T. Noecker, and Anhphuc Phan

6 **Wireless Lan-Based Vehicular Location Information Processing** 69
Seigo Ito and Nobuo Kawaguchi

7 **Perceptually Optimized Packet Scheduling for Robust Real-Time Intervehicle Video Communications** 83
Enrico Masala and Juan Carlos De Martin

8 Machine Learning Systems for Detecting Driver Drowsiness 97
Esra Vural, Müjdat Çetin, Aytül Erçil, Gwen Littlewort, Marian Bartlett, and Javier Movellan

9 Extraction of Pedestrian Regions Using Histogram and Locally Estimated Feature Distribution 111
Kenji Mase, Koji Imaeda, Nobuyuki Shiraki, and Akihiro Watanabe

10 EEG Emotion Recognition System 125
Ma Li, Quek Chai, Teo Kaixiang, Abdul Wahab and Hüseyin Abut

11 Three-Dimensional Ultrasound Imaging in Air for Parking and Pedestrian Protection 137
Marco Moebus and Abdelhak Zoubir

12 A New Method for Evaluating Mental Work Load In *n*-Back Tasks 149
Naoki Shibata and Goro Obinata

13 Estimation of Acoustic Microphone Vocal Tract Parameters from Throat Microphone Recordings 161
Ülkü Çağrı Akargün and Engin Erzin

14 Cross-Probability Model Based on Gmm for Feature Vector Normalization 171
Luis Buera, Antonio Miguel, Eduardo Lleida, Alfonso Ortega, and Óscar Saz

15 Robust Feature Combination for Speech Recognition Using Linear Microphone Array in a Car 187
Yasunari Obuchi and Nobuo Hataoka

16 Prediction of Driving Actions from Driving Signals 197
Toshihiko Itoh, Shinya Yamada, Kazumasa Yamamoto, and Kenji Araki

17 Design of Audio-Visual Interface for Aiding Driver's Voice Commands in Automotive Environment 211
Kihyeon Kim, Changwon Jeon, Junho Park, Seokyeong Jeong, David K. Han, and Hanseok Ko

18 Estimation of High-Variance Vehicular Noise 221
Bowon Lee and Mark Hasegawa-Johnson

19 Feature Compensation Employing Model Combination for Robust In-Vehicle Speech Recognition 233
Wooil Kim and John H. L. Hansen

Index 245

Contributing Authors

Mehmet Abbak, Sabancı University, Istanbul, Turkey
Hüseyin Abut, San Diego State University (Emeritus), San Diego, USA; Sabancı University, Istanbul, Turkey
Batu Akan, Sabancı University, Istanbul, Turkey
Ülkü Çağrı Akargün, Koç University, Istanbul, Turkey
Pongtep Angkititrakul, CRSS, University of Texas at Dallas, Richardson, USA
Kenji Araki, Hokkaido University, Sapporo, Japan
Ali Özgür Argunşah, Sabancı University, Istanbul, Turkey
Marian Bartlett, University of California, San Diego, USA
Deborah Bruce, National Transportation Safety Board (NTSB), Washington, DC, USA
Luis Buera, University of Zaragoza, Zaragoza, Spain
Müjdat Çetin, Sabancı University, Istanbul, Turkey
Quek Chai, Nanyang Technological University, Singapore
Sangjo Choi, CRSS, University of Texas at Dallas, Richardson, USA
Emrecan Çökelek, Sabancı University, Istanbul, Turkey
Serhan Coşar, Sabancı University, Istanbul, Turkey
Tyler Creek, CRSS, University of Texas at Dallas, Richardson, USA
Baran Çürüklü, Karolinska Institutet, Stockholm, Sweden
Serhan Danış, Istanbul Technical University, Istanbul, Turkey
Juan Carlos De Martin, Politecnico di Torino, Torino, Italy
Aytül Erçil, Sabancı University, Istanbul, Turkey
Hakan Erdoğan, Sabancı University, Istanbul, Turkey
Kayhan Eritmen, Sabancı University, Istanbul, Turkey
Engin Erzin, Koç University, Istanbul, Turkey
Rahmi Fıçıcı, Sabancı University, Istanbul, Turkey
David K. Han, Naval Academy, Annapolis, Maryland, USA
John H.L. Hansen, CRSS, University of Texas at Dallas, Richardson, USA
Nobuo Hataoka, Tohoku Institute of Technology, Sendai, Japan
Jeremy Hayes, CRSS, University of Texas at Dallas, Richardson, USA
Koji Imaeda, Nagoya University, Nagoya, Japan
Mümin Imamoğlu, Sabancı University, Istanbul, Turkey

Seigo Ito, Nagoya University, Nagoya, Japan
Toshihiko Itoh, Hokkaido University, Sapporo, Japan
Katsunobu Itou, Hosei University, Tokyo, Japan
Changwon Jeon, Korea University, Seoul, Korea
Seokyeong Jeong, Korea University, Seoul, Korea
Mark Hasegawa-Johnson, University of Illinois at Urbana-Champaign, Urbana, USA
Teo Kaixiang, Nanyang Technological University, Singapore
Harun Karabalkan, Sabancı University, Istanbul, Turkey
Çagatay Karabat, Sabancı University, Istanbul, Turkey
Mehmet Karaca, Sabancı University, Istanbul, Turkey
Nobuo Kawaguchi, Nagoya University, Nagoya, Japan
Jeonghee Kim, CRSS, University of Texas at Dallas, Richardson, USA
Kihyeon Kim, Korea University, Seoul, Korea
Wooil Kim, CRSS, University of Texas at Dallas, Richardson, USA
Norihide Kitaoka, Nagoya University, Nagoya, Japan
Hanseok Ko, Korea University, Seoul, Korea
Hakkı Can Koman, OTAM Research Center, Istanbul Technical University, Ayazağa, Istanbul, Turkey
Takashi Kusakawa, Nagoya University, Nagoya, Japan
Donggu Kwak, CRSS, University of Texas at Dallas, Richardson, USA
Bowon Lee, Hewlett-Packard Laboratories, Palo Alto, CA, USA
Ma Li, Nanyang Technological University, Singapore
Gwen Littlewort, University of California, La Jolla, San Diego, USA
Eduardo Lleida, University of Zaragoza, Zaragoza, Spain
Bruce Magladry, Office of Highway Safety, National Transportation Safety Board, Washington, DC, USA
Enrico Masala, Politecnico di Torino, Torino, Italy
Kenji Mase, Information Technology Center, Nagoya University, Nagoya, Japan
Antonio Miguel, University of Zaragoza, Zaragoza, Spain
Chiyomi Miyajima, Nagoya University, Nagoya, Japan
Marco Moebus, Technische Universität Darmstadt, Darmstadt, Germany
Javier Movellan, University of California, La Jolla, San Diego, USA
Takanori Nishino, Nagoya University, Nagoya, Japan
Levi T. Noecker, CRSS, University of Texas at Dallas, Richardson, USA
Goro Obinata, EcoTopia Science Institute, Nagoya University, Nagoya, Japan
Yasunari Obuchi, Central Research Laboratory, Hitachi Ltd., Japan
Alfonso Ortega, University of Zaragoza, Zaragoza, Spain
Junho Park, Korea University, Seoul, Korea
Anhphuc Phan, CRSS, University of Texas at Dallas, Richardson, USA
Óscar Saz, University of Zaragoza, Zaragoza, Spain
Volkan Sezer, Istanbul Technical University, Istanbul, Turkey

Naoki Shibata, Graduate School of Engineering, Nagoya University, Nagoya, Japan
Nobuyuki Shiraki, Vehicle Safety Center, Toyota Central R&D, Aichi-gun, Japan
Kazuya Takeda, Nagoya University, Nagoya, Japan
Fatih Taş, OTAM Research Center, Istanbul Technical University, Ayazağa, Istanbul, Turkey
Sadayuki Tsugawa, Meijo University, Nagoya, Japan
Mustafa Gökhan Uzunbaş, Sabancı University, Istanbul, Turkey
Esra Vural, Sabancı University, Istanbul, Turkey
Abdul Wahab, Nanyang Technological University, Singapore
Akihiro Watanabe, Vehicle Safety Center, Toyota Central R&D, Aichi-gun, Japan
Shinya Yamada, Hokkaido University, Sapporo, Japan
Kazumasa Yamamoto, Toyohashi University of Technology, Toyohashi, Japan
Abdelhak Zoubir, Technische Universität Darmstadt, Darmstadt, Germany

Introduction

In June 2007, the "Third Biennial Workshop on DSP (digital signal processing) for Mobile and Vehicular Systems" took place in Sait Halim Paşa Yalısı, Istanbul, Turkey, with 33 excellent paper presentations from all over the world and a welcoming speech by Cemil Arıkan, Director of Research and Graduate Studies at Sabancı University, Istanbul, Turkey. It was followed by two keynote addresses: "Role of Intelligence in Driving with Anecdotes from Past and Predictions for Future" by Jan Nahum, CEO of Hexagon, Turkey, and "Improved Vehicle Safety & How Technology Will Get Us There, Hopefully" by Bruce A. Magladry, Director of the Office of Highway Safety, U.S. National Transportation Safety Board (NTSB), Washington, DC, USA. In addition, there were two information-rich plenary talks: "New Concepts on Safe Driver Assistance Systems" by Sadayuki Tsugawa, Meijo University, Nagoya, Japan, and "Human Centric, Holistic Perception for Active Safety," Mohan M. Trivedi, University of California at San Diego, La Jolla, CA, USA.

This meeting, third in a series, was a continuation from two earlier workshops in Nagoya, Japan, April 2003, and in Sesimbra, Portugal, September 2005. With the widespread acceptance and frequent acknowledgement of two books published as offspring of those two workshops, it was decided to put together this book with special emphasis on *international partnership* in data collection and *collaborative research* on driver modeling for improved safety and comfort. After carefully reviewing papers, keynote addresses, and the plenary presentations, 19 works from this unique workshop series were selected and authors were asked to formulate extended book chapters in order to provide a broad coverage of the fields in DSP for mobile and vehicular systems, namely, safety, data collection, driver modeling, telecommunication applications, noise reduction, and speech recognition for in-vehicle systems. The chapters in this book can be broadly categorized into five parts.

First, we have the two chapters from two keynote speakers of the biennial workshop. They are chapters that give an overall view of the safety-related work that has been going on in the community from the viewpoint of a senior administrator on highway safety officer in the US Government and a leading authority on ITS (Intelligent Transportation Systems) in Japan.

The second part consists of Chapters 3, 4, and 5 and they present reports from three collaborative large-scale real-world data collection efforts in Istanbul, Nagoya, and Dallas, respectively, which are performed under an umbrella support by the New Energy and Development Organization (NEDO) in Japan, three national and one European Union 7th Frame grants. These chapters provide detailed accounts of data collection and describe data-sets collected for a joint international project with three legs in these countries.

A small data set from each of these three sites is collected into a DVD called "Drive-Best" (Driver Behaviour Signals for In-Vehicle Technology), which can be obtained directly from the editors. It is intended for the scientific community, the technology developers, and the business concerns as a brief introduction to the signals related to driver behavior, vehicular sensors, and the road ahead in an integrated and multi-modal fashion and to promote more research on the subject.

In this DVD, there are folders with data collected collaboratively in Japan, Turkey, and the USA. Additionally, a folder called Italy has inter-vehicle communication traces (related to work in Chapter 6) and a slide with collage of research activity photos is also included for reference. The complete DVD has been up-loaded to three websites and it is updated from time to time. This material can be downloaded free of charge. The information on how to download can be found at the inside cover of this book.

Chapters 6 and 7 form the third group and are related to wireless communication and networking aspects of driving support systems, one for location finding and another one for vehicle-to-vehicle communications.

Some of the remaining chapters are related to a diverse range of applications related to driving and vehicles. There are two chapters on vision-based techniques for drowsiness prediction (Chapter 8) and pedestrian detection (Chapter 9). In addition, there is a study on EEG-based emotion recognition (Chapter 10), and yet, another one on a 3D air-ultrasound imager for car safety applications (Chapter 11). Next, there is a work on the analysis of a method based on involuntary eye movement information for mental workload determination (Chapter 12) and a work on predicting driver actions using driving signals (Chapter 16).

The fifth and final group of chapters is related to robust automatic speech recognition (ASR) in a vehicular environment. These chapters approach the robustness issue using few sound techniques and analyze different solutions to provide robustness in ASR. The topics range from bone microphones (Chapter 13) or microphone arrays (Chapter 15) to audio–visual information fusion (Chapter 17) for improved performance. In addition some chapters analyze a better normalization method (Chapter 14) or a model combination-based feature compensation approach (Chapter 19) for robustness. Chapter 18 addresses noise estimation which is useful for robust ASR.

We hope this book will provide an up-to-date treatment of vehicular signal processing, with new ideas for researchers and comprehensive set of references for engineers in related fields. We thank all those who participated in the 2007 workshop. We kindly acknowledge support from the NEDO in Japan, the U.S. National Science Foundation (NSF), national and international funding agencies, Sabancı University in Turkey and Nagoya University in Japan for their support in organizing the "Third Biennial Workshop on DSP for Mobile and Vehicular Systems" in Istanbul, Turkey, June 2007. We wish to express our continued appreciation to Springer Publishing for ensuring a smooth and efficient publication process for this textbook. In particular, we are thankful to Alex Greene and Ms. Jennifer Mirski of Springer for their untiring efforts to make this book better and providing us a high-quality and scholarly platform to stimulate public awareness, fundamental research, and technology development in this unique area.

June, 2008

The Editors,
Kazuya Takeda,
John H.L. Hansen,
Hakan Erdoğan, and
Hüseyin Abut

1 Improved Vehicle Safety and How Technology Will Get Us There, Hopefully

Bruce Magladry and Deborah Bruce

Abstract The successful deployment of new technologies in highway vehicles hinges on the driver's ability to safely use those systems. This chapter calls on the engineering community to give full consideration to the usability problems associated with in-vehicle systems designed to engage/communicate with drivers. Such interactions may contain information about the vehicle, the roadway, other vehicles, the route, or the weather, or they may be of personal entertainment interest. There is considerable evidence that drivers are in visual overload, and the delivery of additional information via auditory displays is warranted, but there is cognitive workload associated with driving activities regardless of the perceptual channel involved. The distraction costs for naturalistic speech interaction may be less than for the visual dashboard display of information, but there are many human factors issues to address in order to ensure improved driver performance and safety.

Keywords Highway safety · In-vehicle information systems · Driver behavior · Human factors · Attention · Perception · Distraction · Crash avoidance technology · Hand-held wireless device · Auditory displays · NTSB

1.1 Introduction

The National Transportation Safety Board (NTSB) investigates highway accidents in order to make recommendations to improve highway safety. It is from that perspective that this chapter considers digital signal processing (DSP) for mobile and vehicular systems. Highway safety programs seek to improve safety either by preventing crashes or by increasing crash survivability. US public policy has reached some practical limits in occupant protection and crash mitigation; consequently, new programs, such as intelligent transportation systems (ITS), focus on crash avoidance to improve safety. With a few

B. Magladry (✉)
National Transportation Safety Board (NTSB), Washington, DC, USA

K. Takeda et al. (eds.), *In-Vehicle Corpus and Signal Processing for Driver Behavior*, DOI 10.1007/978-0-387-79582-9_1,

exceptions, like stability control systems, technologies for crash avoidance involve the driver as a critical control element in the system's performance. This chapter looks at the human factors influences on in-vehicle system designs.

1.2 Highway Safety

The NTSB investigates transportation accidents in all modes of travel—highway, aviation, marine, rail, and pipeline. It is important to realize that the Safety Board is independent of the regulators in the US Department of Transportation (USDOT). That arrangement was carefully constructed to ensure that NTSB investigations and NTSB safety recommendations are unbiased. Many governments around the world have similar organizations.

There are many ways to measure safety, but no one can argue that the bottom line is fatalities. With the exception of highway travel, most modes of transportation in the United States experience between 700 and 800 fatalities per year.[1] The number of fatalities in the marine mode each year is approximately 800; the vast majority of those drown in recreational boating accidents. About the same number of fatalities occurs in rail accidents annually, where majority are trespassers and rail workers, not passengers. In aviation, the average is about 750 fatalities each year, almost all associated with private pilots in small general-aviation aircraft. We also have about a dozen pipeline fatalities each year from gas explosions.

By comparison, the United States had approximately 43,300 highway fatalities and 2.5 million injuries from nearly 6 million crashes last year [1]. Every day, more than 16,000 crashes occur on American highways. With a population of about 300 million, we have over 250 million registered vehicles. The automotive industry is a major economic force in the United States, but highway injuries and fatalities are a drag against that economic engine. Motor vehicle crashes in the United States are estimated to cost more than $230 billion per year [2].

1.3 Drivers

DSP development work focuses on many different aspects of in-vehicle information systems, including biometric-based authentication; telematics and the associated interface functionality of speech recognition and interaction; autonomous navigation; and personal communications. The technical realization that we *can* deploy new DSP functionality should always be balanced against the strategic question of "*but should we*?" Yes, we can develop robust system performance under a variety of environmental conditions and at an acceptable

[1] Recent 5-year averages of annual fatalities by transportation mode are marine 774, rail 806, aviation 752, highway 42,731, and pipeline 14.

cost, but whether we should is a value question predicated on a hierarchy of driving tasks. Will the new system directly improve vehicle control? Will it assist in navigation? Will it better inform travelers without negatively impacting driver performance?

The driving environment is defined by many different interactive factors, such as type of vehicle, route, time of day, weather, amount of traffics plus a whole host of activities that go on inside the vehicle—monitoring children, eating, listening to music, making phone calls, etc. Drivers receive a basic licensing test; but they undergo no recurrent training, they receive no medical evaluation, and their education and language skills vary widely. Drivers may be totally inexperienced in their vehicle type, may have conducted no trip planning, and may view driving as secondary to other personal activities in the car. Furthermore, many drivers do not take the time to understand their cars, do not understand how their driving habits affect their safety, and have not read their owner's manuals.

By and large, driving is a simple task; it must be, because nearly everyone is able to do it. Beginning with inexperienced 16-year-olds all the way through 70- to 80-year-old senior citizens, drivers exhibit a wide range of abilities. However, compared to cars of a generation ago, new vehicles with electronic stability control, moving map displays, bluetooth phone connections, iPod-driven stereos, and passenger video displays present an array of complicated control tasks. As we work to improve and integrate those electronic systems and add functionality to the next generation of highway vehicles, we need to cautiously consider the role of the driver. From a safety vantage point, mistakes can be really costly.

1.4 Attention, Perception, and System Interfaces

The Safety Board has a 50-year history of investigating human performance in transportation accidents. During that time, we have benefited from advances in the sciences of human factors and cognitive ergonomics. The influence of human factors engineering throughout the design process has resulted in early prototyping, task mapping, designing for error management, and exploiting the use of system constraints to enhance safety.

A user-centered design philosophy [3] for in-vehicle system development is becoming the norm, rather than the exception. Designers first ask: what problem are we solving? For in-vehicle systems, that should be a driving problem. To do this effectively, we need to test the validity of the assumptions that we incorporate into the design—particularly assumptions about the driver and a wide range of behavioral, cognitive, perceptual, and psychodynamic factors. These individual differences are the very distinctions exploited by DSP driver recognition systems.

Driving is a highly visual task, requiring almost continuous attention while the vehicle is in motion. Yet the proliferation of in-vehicle computer displays has proceeded despite the fact that reference to driver's overload from visual information has existed in the research literature for several decades [4, 5, 6, 7, 8]. Drivers are quite often operating beyond their visual or perceptual capabilities in a number of key driving situations, including when overtaking another vehicle, when joining or crossing high-speed roads, or when responding to a number of nighttime situations.

Given the heavy demand that driving places on visual perception, it is prudent to consider alternative display modalities [9]. Unlike the perceptual channel for visual processing, auditory perception is not overloaded during the driving task. It is, in fact, rarely used [10]. Although auditory display research in the driving domain is somewhat limited, the results are generally positive. Auditory route guidance information has been associated with more efficient driving, as measured by time and distance [11]; auditory route guidance devices result in fewer navigational errors [12]; and drivers have been found to react faster and with fewer errors using auditory information systems instead of visual systems [13]. Deatherage defined a set of guidelines for selecting auditory or visual display channels based on the characteristics of the message, the environment, and the task [14]. Using these guidelines, auditory presentation of information is appropriate when messages (1) are simple, short, and temporal in nature; (2) require immediate action; and (3) do not have to be referred to later.

However, system interface solutions are not as simple as "auditory instead of visual" information display. The workload associated with attending to in-vehicle displays depends on the complexity of the message, the interaction requirements necessary to manipulate the system, and the time pressures associated with processing the information. Simply put, there is a cognitive capacity limit that is independent of the perceptual mode. Even if neither the visual nor the auditory perceptual channel is overloaded, the sum of incoming information can create a cognitive processing bottleneck. The result is slowed reaction time, distraction, and a narrowing of focus referred to as "tunneling" that result in missed information.

Researchers have long found it useful to categorize different types of driving activities [15], and we now rather commonly refer to distinctions between control, guidance, and navigation tasks. For example, driving activities associated with travel trip planning and navigation have an elastic window of time that may or may not affect vehicle control. Many aspects of navigation can be deferred until the traffic situation avails an opportunity to consider the information being presented, thereby avoiding the cognitive load of multiple, concurrent activities. However, from a design point of view, it is preferable to constrain the system in ways that do not call on the driver to assign a real-time hierarchy to cognitive demands.

Unlike strategic planning tasks, a different situation exists for in-vehicle systems designed to augment real-time operation and control of the vehicle. These systems, which provide the driver with information concerning traffic signs, the direction of the next turn, and collision avoidance, are time-critical because they focus on vehicle control activities that have a finite timeframe of performance. Time constraints are an important characteristic to distinguish between strategic tasks such as route planning and tactical tasks such as lane tracking [16]. Theoretical research into attention mechanisms and applied research into driver performance indicate that in-vehicle display systems should minimize time-dependent, dual-task situations. This in turn means that temporal distinctions between auditory and visual display should be factored into systems designed to support driving tasks.

1.5 In-Vehicle System Technologies

It is also useful to distinguish between vehicle systems for crash avoidance, which become integral to the operation of the car and often demand very little interaction on the part the driver, and those systems that incorporate the driver as a control loop component. Many manufacturers already offer some form of crash avoidance technology on current car models. These autonomous systems affect stability control, rollovers, lane departures, and rear-end/forward collisions without relying on driver-initiated behavior.

A different category of systems focuses on communicating with the driver. For example, vehicle-centered services, such as remote diagnostics, remote vehicle access, weather sensing, and collision notification, are currently available on many cars. Already, commercial fleet operators use data communications to contact mobile fleets, issue preplanned routes, and notify drivers of scheduled maintenance. In the future, such transmissions will include vehicle software upgrades, malfunction and diagnostic reports, and the capability to order parts and receive recall and service notifications.

We will eventually see basic connectivity for the life of the vehicle without the need for ongoing subscription payments, working through a shared message handling utility on behalf of all manufacturers. Highway information for drivers will be as affordable and common as FM radio: the broadcast spectrum for this technology has been identified, and geo-stationary satellites and ground-based towers are planned for 2012, with limited rollout by 2009.

Integrated vehicle-based safety systems is a new USDOT vehicle safety initiative to build and field-test integrated crash warning systems to prevent rear-end, lane change, and roadway departure collisions on light vehicles and heavy commercial trucks. These systems are being deployed in cars as well.

1.6 Concerns for Driver Distraction

Human factors engineers know that effective interfaces begin with an analysis of what the person is trying to do, rather than as a metaphor for what the system should display. This distinction between merely providing information and helping with the activity is at the heart of facilitating driver performance rather than causing driver distraction.

A recent naturalistic driving study funded by NHTSA and conducted by the Virginia Tech Transportation Institute is yielding a host of real-world data about driver behavior [17]. That large-scale, instrumented vehicle study included approximately 43,000 h of data on 241 drivers traveling 2 million miles. Early results indicate that visual inattention and engaging in secondary tasks contribute to 60% of crashes. More specifically, looking away from the forward roadway for more than 2 s doubles the odds of being in a crash. Evidence clearly suggests that tasks requiring longer and more frequent glances away from the road are detrimental to safe driving [18]. It should also be noted that even though police reports most certainly understate the magnitude of the problem, inattention has been cited in one-third of all rear-end and lane-change crashes [19].

The 100-car study concluded that dialing a hand-held wireless device increases risk by a factor of 2.8×, but the risk was not just associated with manipulating the phone; talking or listening on hands-free devices increased the risk by one-third (1.3×). This is consistent with earlier research. In 2005, the Insurance Institute for Highway Safety found that Australian drivers using cell phones were four times as likely to be involved in a serious crash as non-users, regardless of whether they used hands-free devices like earpieces or speaker phones.

If we think about in-vehicle auditory displays as talking boxes, continually telling the driver what is visually obvious, then they are synonymous with a nuisance display. Instead, we need to design auditory displays that focus on intelligent selection of information appropriate for the driving task. The future car may be able to acquire information about all 54 million roadside signs, but the driver only needs to know what a sign says when it would affect the safe control of their vehicle or when they specifically request a category of information. Drivers only need to be informed of railroad crossings when trains are active at that crossing. Drivers need not be informed of every cross street along their route of travel, just the one associated with their next turn.

System integration is also an important issue. Different manufacturers make anti-lock brakes, stability control systems, collision avoidance—and these systems must work in concert to avoid a variety of road hazards. Developers of these technologies must consider how the systems will be used, where displays will be located, how much information is needed, what information has priority, when the systems should be active, and how the systems should function in an emergency.

1.7 Conclusions

Engineering research has made major advances in in-vehicle driver engagement/communication capabilities, but the human factors aspects of the driver interface will be critical to successful implementation of these systems. Drivers, the information that they use, and the environmental conditions in which they operate are necessary components for directly evaluating the suitability of in-vehicle information systems. The functionality of in-vehicle devices should be obvious; drivers should not interact with the information technology device in their car, rather, the technology should seamlessly assist them with driving tasks. Evolution of the computer interface should lead to, as Brenda Laurel calls it, "direct engagement" [20]. In the end, it is the public, and their ability and willingness to make use of these systems, that will determine how effective they will be—and how soon. Ultimately technology is migrating into the vehicle; it is up to those developing these technologies to ensure these advances contribute to improved comfort, usability, and safety.

References

1. National Highway Safety Administration (NHTSA) 2006 preliminary data, per USDOT NHTSA press release dated June 1, 2007.
2. U.S. Department of Transportation, National Highway Transportation Safety Administration, *The Economic Impact of U.S. Motor Vehicle Crashes, 2000*, DOT-HS-809-446 (Washington, DC: NHTSA, 2002).
3. Norman, D., *The Psychology of Everyday Things* (New York: Basic Books, 1988).
4. Sivak, M., The information that drivers use: is it indeed 90% visual? *Perception*, 25, pp. 1081–1089 (1996).
5. Dewar, R.E., In-vehicle information and driver overload. *International Journal of Vehicle Design*, 9, pp. 557–564 (1988).
6. Zwahlen, H.T., & DeBald, D.P., Safety aspects of sophisticated in-vehicle information displays and controls. *Proceedings of the 30th Annual Meeting of Human Factors and Ergonomics Society*, pp. 604–609 (1986).
7. Senders, J.W., Kristofferson, A.B., Levinson, W.H., Dietrich, C.W., & Ward, J.L., The attentional demand of automobile driving. *Highway Research Record*, 195, pp. 15–33 (1967).
8. Treat, J.R., Tumbas, N.S., McDonald, D.T., Shinar, D., Hume, R.D., Mayer, R.D., Stansifer, R.L., & Castallen, N.J., *Tri-level Study of the Cause of Traffic Accidents*. Report No. DOT-HS-034-3-535-77, Department of Transportation (1977).
9. Mollenhauer, M., Lee, J., Cho, K., Hulse, M., & Dingus, T., The effects of sensory modality and information priority on in-vehicle signing and information systems. *Proceedings of the 38th Annual Meeting of Human Factors and Ergonomics Society*, pp. 1072–1077 (1994).
10. McKnight, A.J., & Adams, B.B., *Driver Education Task Analysis, Volume 1: Task Description Report*. DOT HS 800 367, Department of Transportation (1970).
11. Streeter, L.A., Vitello, D., & Wonsiewicz, S.A., How to tell people where to go: comparing navigation aids. *International Journal of Man-Machine Studies*, 22, pp. 549–562 (1985).
12. Walker, J., Alicandri, E., Sedney, C., & Roberts, K., *In-Vehicle Navigation Devices: Effects on the Safety of Driving Performance*. Tech Report FHWA-RD-90-053, Department of Transportation (1990).
13. Srinivansan, R., & Jovanis, P.P., Effects of selected in-vehicle route guidance systems on driver reaction times. *Human Factors*, 39(2), pp. 200–215 (1997).
14. B.H. Deatherage, "Auditory and other sensory forms of information presentation," eds. H.P. Van Cott and R.G. Kinkade, *Human Engineering Guide to Equipment Design* (Washington, DC: U.S. Government Printing Office, 1972).
15. Lunenfeld, H., & Alexander, G.J., Human factors in highway design and traffic operations. *Journal of Transportation Engineering*, 110(2), p. 150 (March 1984).

16. Michon, J.A., "A critical view of driver behavior models: what do we know, what should we do?," L. Evans and R.C. Schwing (eds.) *Human Behavior and Traffic Safety* (New York: Plenum Press, 1985).
17. Dingus, K., Neale, P., Lee, S., et al., *The 100-Car Naturalistic Driving Study*, 2006.
18. U.S. Department of Transportation, National Highway Transportation Safety Administration, *The Impact of Driver Inattention on Near-Crash/Crash Risk: An Analysis Using the 100-Car Naturalistic Driving Study Data*, DOT-HS-810-594 (Washington, DC: NHTSA, 2006).
19. U.S. Department of Transportation, National Highway Transportation Safety Administration, *Examination of Crash Contribution Factors Using National Crash Databases*, DOT-HS-809-664 (Washington, DC: NHTSA, 2003).
20. B. Laurel, *Computers as Theatre* (Boston, MA: Addison-Wesley Professional, 1993).

2
New Concepts on Safe Driver-Assistance Systems

Sadayuki Tsugawa

Abstract This chapter introduces new concepts on safe driver-assistance systems for increase in driver acceptance of the systems and for elderly driver assistance. Driver acceptance will be increased by helpful but not- annoying assistance, which requires monitoring of the driver status as well as the traffic and roadway conditions. The assistance has some levels like information, or advice, or warning, which is determined by both the driver status and the traffic and roadway conditions. The elderly driver-assistance is based on cooperation between vehicles, which assists not only perception but also decision making and action by an elderly driver from another vehicle through inter-vehicle communications. Some experiments of the systems will also be described.

Keywords Intelligent transport systems (ITS) · Driver assistance · Navigation systems · Elderly drivers · Lidar · Lane keeping · Collision mitigation · Steering assistance · Obstacle detection · Adaptive cruise control (ACC) · Driver-adaptive systems · Cooperating vehicles · Dedicated short-range communication (DSRC)

2.1 Introduction

In recent years intelligent transport systems (ITS) activities in the world have mainly focused on road traffic safety. The research and development of safe driver-assistance systems has been actively conducted by both governments and industries in Japan, Europe and the US, like AHS (advanced cruise assist highway system) and ASV (advanced safety vehicles) in Japan, PReVENT in EU, and VII (vehicle infrastructure integration) in the US [1].

There are still many issues facing such systems, one being the market introduction and deployment of the driver-assistance systems. Although national

S. Tsugawa (✉)
Department of Information Engineering, Meijo University, Shiogamaguchi 1-501, Tempaku-ku, Nagoya-shi 468-8502, Japan
e-mail: tsugawa@ccmfs.meijo-u.ac.jp

K. Takeda et al. (eds.), *In-Vehicle Corpus and Signal Processing for Driver Behavior*, DOI 10.1007/978-0-387-79582-9_2,

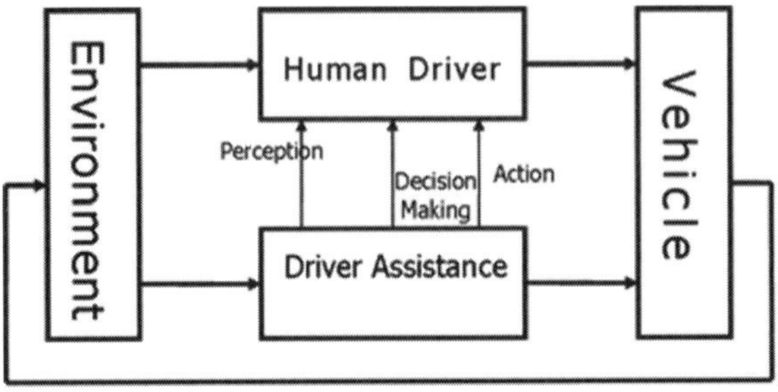

Fig. 2.1 Configuration of a driver-assistance system

projects on safe driver-assistance systems have been conducted and some of the results have been commercialized and put on the market, at present the penetration rate of driver-assistance systems in Japan is quite low. For example, more than 2 million car navigation systems have been put on the market every year, which indicates about half of new cars are equipped with it, but only a few driver-assistance systems compared with the car navigation systems have been put on the market every year; for example, in 2004, curvature warning systems that sold the most among commercially available safe driver-assistance systems were equipped in 3% of new cars. The reasons of the low penetration rate are the cost and the driver acceptance.

Another issue, which may be peculiar to Japan, is elderly driver assistance. At present the rate of elderly drivers (65 years old or more) is 10% of all the drivers, and it will increase to more than 15% in the next 10 years. The number of fatal accidents caused by elderly drivers has tripled in the past 17 years.

This chapter presents new concepts of the safe driver-assistance systems for increase in driver acceptance of the systems and for elderly driver assistance. A driver-assistance system can be defined as a system that helps a driver in perception, or decision making, or action while driving, as shown in Fig. 2.1. Since a driver is a time-variant system, the assistance should be adaptive to a driver, which requires sensing or monitoring of a driver status. This concept is the first topic studied in this chapter. When it is difficult to assist a driver with onboard equipment, the driver should be assisted from the outside of a vehicle, which is another focus of this chapter.

2.2 Driver-Assistance Systems in the Market

Before introducing new concepts on driver-assistance systems, current driver-assistance systems on the market will be introduced. Driver-assistance systems, as shown in Fig. 2.1, are defined as those in which a mechanism or system covers part of the sequence of tasks (recognition, decision making, action) that a driver performs while driving a motor vehicle. Mechanisms or systems that conduct all such tasks are known as automated driving systems. Driver-assistance systems

Table 2.1 Driver-assistance systems on the market

Year	Systems	Makers
1989	Forward collision warning with lidar	Nissan Diesel
1995	ACC	Mitsubishi
2001	Lane-keeping support	Nissan
2003	Collision mitigation brakes	Honda
2004	Low-speed ACC	Nissan
2004	Night vision	Honda

are characterized by the inclusion of the human driver in the control loop, making the human–machine interface of critical importance. Today, all driver-assistance systems that have been commercialized are stand-alone, single-vehicle systems. Here, we introduce single-vehicle driver-assistance systems that have been brought onto the market. Most of these systems are marketed not as safe driver-assistance systems but as comfort or convenience systems. Table 2.1 presents a timeline when major driver-assistance systems have been commercialized.

Lateral Assistance and Control Systems: Lateral assistance and control systems (which support and control side-to-side vehicle movement, i.e., steering) include both systems that provide steering assistance when driving at high speed and those that support automated steering while parking. Note that power steering, being a handling aid, is not generally considered as a driver-assistance system.

Steering Assistance: Lane-keeping support systems that use machine vision to detect lane markers (white lines) on the road in front of the vehicle and prevent lane departure have been commercialized domestically for passenger vehicles. Such systems provide assistance by applying a slight torque to keep the vehicle traveling in the lane, but their human–machine interface varies by product. In one, a driver who tries to steer in the wrong direction feels a counter torque [2]. In another, the system provides assistance by applying a slight torque as the driver continues to steer; should the driver discontinue steering for a period of 5 s, the system will sound a caution and terminate assistance [5]. In either case, as mentioned above, the products are positioned not as lane departure warning systems but as systems for improving comfort.

Systems Supporting Automated Steering During Parking: Systems have been brought to market that use a camera installed near the rear license plate that support automatic steering when backing up to park [4]. The camera is used when the driver sets the parking location within the parking space displayed on the screen of the car navigation system prior to initiating parking, not to provide feedback to the system while the vehicle is backing up. That is, parking-assistance system involves open-loop control of traveling distance and automatic steering to the location set by the driver. Therefore, if the wheels slip it may no longer be possible to park in the target location. Electric power steering is used as an actuator in steering.

Longitudinal Assistance and Control Systems: Here we introduce longitudinal assistance and control systems (which support and control front-to-back vehicle movement, i.e., assistance and control acceleration and braking) including obstacle detection in the surrounding area, adaptive cruise control (ACC) that supports and controls both speed and inter-vehicle distance, and collision mitigation brakes. Other systems that have been brought to market include brake assistance systems that supplement the braking force applied by the driver, anti-lock brake systems (ABS) that prevent wheel-lock and the loss of steering control caused by braking on surfaces with a low coefficient of friction, and vehicle stability control systems that maintain vehicle stability and enable proper steering and braking on surfaces with a low coefficient of friction.

Recognition and Decision Assistance: Autonomous recognition-assistance systems that have been put onto the market include

- Systems that display the location of obstacles detected using ultrasound when backing up
- Systems that support recognition when backing up by displaying images, captured by a camera installed near the rear license plate, on a car navigation system while backing up
- Proximity warning systems that use caution sounds and displays to notify the driver of obstacles in the surrounding area as detected by ultrasound during low-speed driving

Such systems support recognition by identifying the presence of obstacles behind the vehicle when backing up and around the vehicle when driving at low speed, but it is the driver who confirms safe conditions. Systems that detect and display humans and animals forward of the vehicle have also been commercialized. These include both passive systems, which detect the infrared rays emitted by subjects and convert the infrared image to a visible image for display, and active systems, which emit infrared rays from the vehicle and then convert the reflected light to a visible image for display. These systems go no further than the stage of recognition assistance, but systems that use infrared images to detect pedestrians and display them to drivers have been commercialized [7], which support not only recognition but also decision making.

Adaptive Cruise Control: Adaptive cruise control (ACC) goes beyond conventional cruise control by, in the presence of a leading vehicle, automatically controlling vehicle speed in accordance with the distance from and speed relative to the leading vehicle. Measurement of inter-vehicle distance is often performed using lidar, although some systems on the market employ millimeter-wave radar. Inter-vehicle distance can generally be set at three levels: far, medium, and near. Lidar is superior to millimeter-wave radar, discussed later, in terms of directional resolution, but is difficult to use when it is raining because of the

Table 2.2 Penetration of driver-assistance systems

System	2000	2001	2002	2003	2004
Curve warning	8,108	10,720	10,335	81,295	133,487
ACC	3,389	6,916	24,102	8,008	17,611
Lane-keeping support		947	422	2,582	1,671
Navigation-coordinated shift control	193	203	192	31,869	47,884
Drowsiness warning	8,032	10,737	48,334	81,382	124,748
Night vision			407	673	1,923
Collision mitigation brakes				5,244	10,921
Domestic vehicle shipment (x 1,000)	4,576	4,457	4,473	4,435	4,574

Source: Ministry of Land, Infrastructure, and Transport, (2005)

attenuation of the laser light; ACC systems that use lidar are designed not to function when the windshield wipers are in use. Millimeter-wave radar, meanwhile, has a slightly lower directional resolution than lidar, but can be engaged when it is raining. Generally speaking, millimeter-wave radar ACC is found in high-end luxury cars.

Collision Mitigation Brake Systems: These have been brought to market the detect forward vehicles and automatically invoke the brakes to reduce rear-end collisions when the collision cannot be avoided [3]. Forward vehicles are detected using millimeter-wave radar, which adapts well to the environment, at an angle of 16° to the left and the right to cope with straight and gently curving roadways. Onboard sensors including wheel speed sensors that detect the speed of ego vehicle and yaw rate sensors that measure degree of rotation are used to estimate the course of ego vehicle and derive the estimated degree of lateral movement at the location of the obstacle.

System Penetration: Most of the driver-assistance systems mentioned above were developed under the advanced safety vehicle (ASV) project described below. Table 2.2 shows their levels of penetration from the year 2000 to 2004. More than four million new cars are shipped every year in Japan but the penetration rate is extremely low, with even the most frequently adopted systems installed on only about 3% of vehicles. Incidentally, roughly half of new cars are now equipped with car navigation systems, resulting in an accumulated total of more than 20 million systems shipped. One reason for the low penetration of driver-assistance systems is their price.

2.3 Driver-Adaptive Assistance Systems

One requirement to increase the driver acceptance of a driver-assistance system will be that the system will provide helpful and kind but not- annoying assistance. Such assistance will require a driver monitoring function to assist the driver depending on situations. This assistance system is called a driver-adaptive assistance system. A system named "human centered ITS view aid system" will be introduced [6].

The system includes driver sensing systems for eye closing and eye casting as the driver monitoring devices. The driver monitoring system together with other sensing systems of the road surface wet condition and the inter-vehicle spacing, the inter-vehicle communications system, and an adaptive display system constitutes the driver-adaptive assistance system.

The main feature of the system is that it is a stand-alone system and does not require any infrastructure intelligence. When a single vehicle is equipped with the sensing systems and the display system, the assistance system works, even if the inter-vehicle communications are not available. If the inter-vehicle communications are available, the functions of the system will be enlarged, because the driver will have information that is gathered and transmitted from another vehicle.

Driver Monitoring: This system includes the eye closing monitoring system for drowsiness detection [7] and the eye casting monitoring system for detection of taking driver's eyes off the road [8].

A single CCD camera is attached at the rear view mirror and is shared by both the eye closing monitoring system and the eye casting monitoring system. The features of the system are that the face of the driver is illuminated by pulsed lights from infrared LEDs, and the LEDs as well as the CCD camera are mounted in the rear view mirror behind the mirror. Since the mirror is transparent for near-infrared lights, but is reflective for visible lights, a driver cannot see the camera, but the camera can see the driver as an infrared image. This configuration makes the vision system robust against the light conditions and, in addition, does not make a driver feel to be monitored with a CCD camera. After the CCD camera captures the face of a driver, the image processing algorithm will extract the part of an eye of the driver to see the eye closing and the eye casting.

Other Sensing Systems: Other onboard sensing systems are the inter-vehicle spacing measurement system and the road surface monitoring system. The inter-vehicle spacing will be measured with a lidar used for the ACC (adaptive cruise control). The spacing data will be used for the warning on tailgating and collision in this system. The road surface monitoring is for the detection of the wet condition of the road surface and is based on the polarized light from the road surface detected and processed with a vision system [9]. When the road surface is wet, the reflected energy of the horizontal component of the polarized light and that of the vertical component differ. The ratio between the vertical component and the horizontal component indicates the existence of water on the road surface.

Sensing data of the road surface and the spacing can be transmitted to following vehicles over the inter-vehicle communications with 5.8 GHz DSRC (dedicated short-range communications).

Driver-Adaptive Display of Information and Warning: Another feature of "human centered ITS view aid system" is that it provides information to a driver in a driver-adaptive manner in both visual and auditory ways not to intervene the task of the driver [10]. The display of the car navigation system is used as the information display device in the system. The contents of the

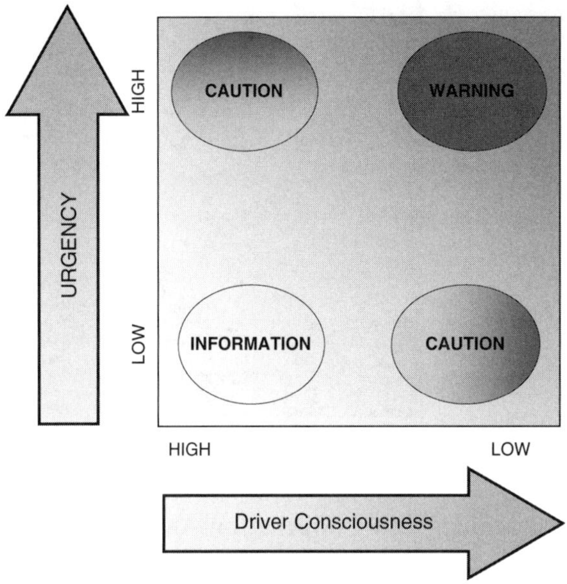

Fig. 2.2 Principle of driver-assistance display

information are determined by the consciousness level of the driver, and the emergency level ahead of the vehicle. Figure 2.2 shows the principle of the display. If the consciousness of a driver, which is detected by the driver monitoring systems, is high and the inter-vehicle spacing is large, the situation is safe and, thus, the display shows no warning. If the consciousness becomes lower or the spacing becomes shorter, the situation is becoming dangerous and, thus, the display shows "caution" and then "danger." The thresholds are determined by the time-to-collision as well as by the consciousness of a driver and can also be adjusted by a driver. The voice messages and the beep sound are also used for the warning.

Experiments: Systems were evaluated with two vehicles equipped with the system. Figure 2.3 shows scenes of the experiments: the road surface wet condition is detected with the onboard sensing system and it is displayed on the display unit of the car navigation unit, and a parked car transmits the situation over the inter-vehicle communications and a following car receives it and displays the information. The system was highly evaluated by experts during demonstration on a test track.

Discussion: Although the issue on the cost of safety devices, which is one main reason why a driver-assistance system does not sell, still exists in the driver-adaptive driver assistance, the idea and concept of "being adaptive to any driver" are highly evaluated by experts [6] through experiments. Since a driver monitoring device has also been put on the market, there will be no high barrier on the cost against driver monitoring for driver-adaptive assistance any more. The driver-adaptive concept is also applicable to elderly driver assistance described below.

Fig. 2.3 One scene from experiments: wet road surface is detected and the information is displayed on the car navigation unit display

2.4 Driver Assistance with Cooperation Among Vehciles

Automobiles are the optimal transportation means for the elderly, because they can provide door-to-door transportation. As the rate of elderly people is rapidly increasing in Japan, the number of elderly drivers is also increasing. The rate of elderly people (above 65 years) exceeded 20% of Japanese whole population in 2005, and it is expected to be 25% in 2015. As a result, the elderly driver population will also be increasing, and it is expected to be 14 millions in 2010, although it was 7 millions in 2002.

As the number of elderly drivers is increasing, the number of traffic accidents caused by elderly drivers is also increasing. The current number of fatal accidents caused by elderly drivers has tripled in the past 17 years. It is the reason why elderly driver assistance will be necessary. This section focuses on a new concept of an elderly driver assistance system.

Concept of the Assistance: An ordinary driver-assistance system generally involves a single vehicle, and sometimes there are bounds on the assistance. An example that shows the existence of the bounds is assistance at a blind intersection. Another example is the assistance on lane changing, and the system can tell the timing of the lane changing, but if the driver is hesitating and cannot make decision to perform lane changing, the driver will have to wait for a gap large enough for the driver to make lane changing.

Driver assistance proposed here is a new one that does not have such bounds found in a driver assistance involving only a single vehicle, but it is based on cooperation between two vehicles, named cooperative driver assistance [11]. In the assistance system, the vehicle that is assisted is called a guest vehicle, and the vehicle that assists is called a host vehicle or an escort vehicle. The host vehicle, usually driven by an experienced driver, will assist or escort the guest vehicle, usually driven by an elderly driver.

Characteristics of Elderly Drivers: Within the framework of the aging society in Japan, there are many studies focusing on the behaviors of elderly people, and the characteristics of the elderly drivers are summarized as follows [12, 13]:

1. They often drive so slowly that they cannot follow the tough traffic flow.
2. They often neglect the stop signs, and it is one of the major causations of fatal accidents by elderly drivers.
3. They often make shortcut on right turning and detour on left turning (in Japan the traffic rules are "keep left").
4. They often make abrupt steering operation.
5. The longitudinal operation of a vehicle by elderly drivers becomes unstable, when the driving workload increases.
6. They often neither care about nor watch their surroundings.
7. They often neglect guidance sign boards.
8. The level of the surroundings watching for safety by elderly drivers is low. In particular, the surrounding watching for safety at intersections is insufficient.
9. The elderly driver ability of the perception of predictable danger and potential danger is low.
10. The self-evaluation of the driving capability by elderly driver is high, but the evaluation by instructors in driving schools is low, which indicates the overconfidence on the driving capability of the elderly drivers.

The characteristics indicate that ordinary driver assistance that uses equipment on a single vehicle will not be sufficient but driver assistance based on a new concept will be necessary.

Principle of Assistance: Characteristics of the elderly drivers above-described indicate that the driver assistance on a single-seat vehicle for an elderly driver must not only be on the single vehicle itself, but also be from another vehicle or the host vehicle. The cooperative driver assistance will be effective especially to elderly drivers, whose driving performance is decreasing. The concept of the system is shown in Fig. 2.4, and the system configuration is in Fig. 2.5. As

Fig. 2.4 Concept of the assistance

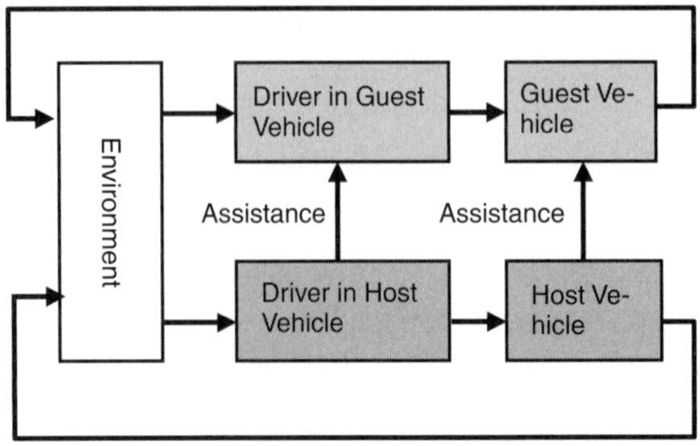

Fig. 2.5 Assistance configuration

illustrated in Fig. 2.4, the fundamental concept of the new assistance is that a vehicle driven by an elderly driver is assisted by another vehicle driven by an experienced driver. We call such assistance "cooperative driver assistance."

Cooperative Driver Assistance: The host vehicle and guest vehicle are provided with the cooperative driver-assistance system in addition to the ordinary, stand-alone driver-assistance system for each vehicle itself. The key technology for the cooperative driver-assistance system is the inter-vehicle communications [14, 15]. Because the cooperative driver assistance must exchange driving data including the localization and the velocity between the vehicles in order to assist and to be assisted, the inter-vehicle communications are essential. In addition, the localization function and the driving instruction function will be necessary.

We have developed a couple of the cooperative driver-assistance systems: a cooperative driver-assistance system at an intersection from an oncoming vehicle, which will be described here in detail; automated parking that employs roadside equipment, mobile or fixed, to guide a vehicle to a parking space [16]; and a two-vehicle platoon, where the lead vehicle is driven by an experienced driver and the trail vehicle, on which an elderly driver is, automatically follows.

Experiments and Evaluation of the Assistance: Experiments on a cooperative driver-assistance system at an intersection by subjects of elderly drivers will be described in detail. The experiments with two experimental vehicles driven by the subjects were conducted along virtual, urban streets constructed on a test track. A traffic signal system was installed at an intersection of the streets.

The assistance system for the experiments and evaluation by elderly drivers is the assistance on the right-turning collision avoidance. The assistance system prevents the crash on the right turning or at the blind intersection, as shown in Fig. 2.6. It provides the information on the vehicles approaching the intersection, including the localization, the speed, and the behavior intention at the intersection (go straight or make turning) with the inter-vehicle communications. It is also possible to provide the information services, warnings, or

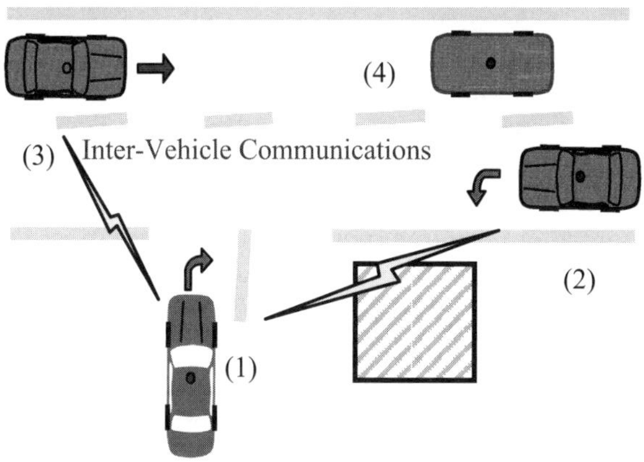

Fig. 2.6 Example for the assistance at a blind intersection

braking control to the driver depending on the stage of the danger. The vehicle (1) is making a right turning, and its intention is transmitted to the vehicles (2) and (3). The vehicle (2) is making a left turning, and its intention is transmitted to the vehicle (1). As a result, the vehicle (1) will make a right turning after the vehicle (3) when it does not yield or before it when it yields.

Vehicles have the functions of the inter-vehicle communications, the localization, and an HMI unit for the instruction to a driver. Each experimental vehicle is equipped with a RTK-GPS receiver, an onboard computer, the map database, the HMI (human–machine interface) units besides an inter-vehicle communication unit. The communication medium is 5.8 GHz DSRC, and the protocol is based on the CSMA (carrier sense multiple access) protocol.

The experiments have been conducted with a newly developed single-seat electric vehicle (Fig. 2.7) and a passenger along virtual streets on a test track

Fig. 2.7 Recently developed single-seat electric vehicle

with the subjects of elderly drivers for the verification of the feasibility and effectiveness of the assistance.

A scenario for an experiment on assistance to an elderly driver making right turning will be as follows. At first, an elderly driver is going to make right turning, and the information is transmitted to the oncoming vehicle. Then, an assistance system on the right-turning vehicle receives information from the oncoming vehicle. The information is output in an auditory form: "A car is coming from the oncoming lane." On the other hand, on the vehicle which an ordinary driver is driving and is going straight, the information is output in an oral form: "An elderly driver is waiting to make right turning on the oncoming lane." Then, the ordinary driver on the going-straight vehicle reduces the speed to allow the right-turning vehicle to make right turning. Finally, on the going-straight vehicle, the driver hears "We yield the way to the right turning vehicle," and on the right-turning vehicle, the elderly driver hears "After you. Go ahead, please." On the other hand, if an ordinary driver on the oncoming vehicle does not reduce the speed, an assistance system on the right-turning vehicle warns the elderly driver not to make right turning. A scene of the experiments with a right-turning vehicle and a going-straight vehicle at the virtual streets on a test track is shown in Fig. 2.8.

The experiment was conducted with 30 subjects of elderly drivers. After the experiments, inquiries on the system were conducted to the subjects for the evaluation of the system. The concept of the cooperative driver-assistance system is well recognized by almost all of them. The response from the subjects shows that over 80% of the elderly drivers positively answered. Some comments on the assistance from the subjects are as follows:

1. Since it is difficult to estimate the speed of an oncoming vehicle, such an assistance system will be helpful for elderly drivers.
2. Although the system may neglect the right of way and may make the traffic rules ambiguous, it is kind and helpful to provide the priority to elderly drivers.

Fig. 2.8 One scene from experiments: vehicle on *left* is making right turning, and another vehicle at the *right* going straight will yield to the left vehicle

2.5 Discussion

The system proposed here assumes that every vehicle has the assistance functions. The assumption is not realistic under general road traffic situations, but the system will be applied to a closed community like an elderly people village or rural areas where the population is sparse and the rate of elderly people is high.

Although the fundamental experiment showed the effectiveness of the system, the following issues still remain, which must be examined in the future:

1. Countermeasures to mixture of vehicles with communication function and vehicles without communication function
2. Identification of the receiver
3. The accuracy and reliability of the localization
4. Distrust and overconfidence in the system
5. The method and timing of the presentation of the assistance information

In terms of systems with the inter-vehicle communications, a system that assumes every vehicle is equipped with a communication device is not feasible. A system that functions even at a low penetration rate of the communication unit must be found.

2.6 Conclusion

This chapter has introduced two new concepts of driver-assistance systems for the increase in the driver acceptance and for elderly driver-assistance. One requirement of the increase in the acceptance of a driver-assistance system is that the system provides helpful and kind, but not annoying assistance to any driver. The driver-adaptive assistance system presented here has a function of driver monitoring as well as traffic and roadway monitoring to provide such assistance.

Since driver assistance with stand-alone, onboard equipment is not sufficient for elderly drivers, a new concept of a driver-assistance system consisting of two vehicles has been presented. Fundamental experiments with subjects of elderly drivers have been conducted to show the feasibility of the assistance at an intersection.

The future issues for the proposals include human–machine interface (HMI) or the detailed design of the driver-adaptive display in auditory, visual, or haptic means. Unlike automated driving, driver assistance requires an excellent HMI for better acceptance and better performance.

Acknowledgments The author would like to thank the members of the consortium "Human Centered ITS View Aid System" (FY2001–FY2003) and NEDO (New Energy and Industrial Technology Development Organization) for the financial support for the elderly driver assistance (FY2002–FY2006).

References

1. S. Tsugawa: Trends and Issues in Safe Driver Assistance Systems: Driver Acceptance and Assistance for Elderly Drivers, IATSS Research, Vol. 30, No. 2, pp. 6–18 (2006).
2. N. Chiku et al.: A Study of Lane Keep Support System Based on Human Driving Characteristics. Proceedings of JSAE Annual Spring Congress 2001, No. 19–01, pp. 1–4 (2001) (in Japanese).
3. S. Ishida et al.: The Method of a Driver Assistance System and Analysis of a Driver's Behavior. Proceedings of the 10th World Congress of ITS (CD-ROM) (2003).
4. T. Endo, et al.: Development of Reverse Parking Assist with Automatic Steering, Proceedings of The 10th ITS World Congress (CD-ROM) (2003).
5. T. Tsuji, et al.: Development of Night Vision System, Transactions of Automative Engineers of Japan, Vol. 33, No. 3, pp. 173–178 (in Japanese).
6. S. Tsugawa et al.: A Driver-Adaptive Driver Assistance System, Proceedings of 3rd ITS Symposium, pp. 409–414 (2004) (in Japanese).
7. H. Mizuno et al.: Measurement of the Driver's Blinking and Estimation of Their Awareness Using the Image Processing, Proceedings of the 9th World Congress of ITS, Paper #3136 (CD-ROM) (2002).
8. S. Mita et al.: Detection System of Gaze Direction for Excessive Concentration Through Cellular Phone Use and Inattention, Proceedings of the 9th World Congress of ITS, Paper #3046 (CD-ROM) (2002).
9. M. Yamada et al.: Study of a Road Surface Condition Detection Technique in the Human Centered ITS View Aid System, Proceedings of the 9th World Congress of ITS, Paper #3063 (CD-ROM) (2002).
10. H. Ishigaki et al.: The Display Method of Route Information and Driver Support Information System Depending on the Degrees of the Consciousness of a Driver and Feature for the Elderly, Proceedings of the 9th World Congress of ITS, Paper #3085 (CD-ROM) (2002).
11. S. Kato et al.: Driver Assistance with Cooperative Driving, Proceedings of AVEC 2004 (CD-ROM) (2004).
12. M. Kamata et al.: Mobility of Elderly People, Proceedings of ITS Symposium, pp. 11–16 (2003) (in Japanese).
13. K. Renge: Elderly Drivers' Characteristics, Proceedings of ITS Symposium, pp. 17–23 (2003) (in Japanese).
14. S. Kato et al.: Applications of Inter-vehicle Communications to Driver Assistance and Automated Platooning, Proceedings FISITA 2002, F02I274 (2002).
15. S. Kato et al.: Applications of Inter-vehicle Communications to Driver Assistance System, JSAE Review, Vol. 24, pp. 9–15 (2003).
16. N. Hashimoto et al.: Autonomous Vehicle Guidance Through Cooperation Between Vehicles, Proceedings of the 13th World Congress of ITS (CD-ROM) (2006).

3
Real-World Data Collection with "UYANIK"

Hüseyin Abut, Hakan Erdoğan, Aytül Erçil, Baran Çürüklü, Hakkı Can Koman, Fatih Taş, Ali Özgür Argunşah, Serhan Coşar, Batu Akan, Harun Karabalkan, Emrecan Çökelek, Rahmi Fıçıcı, Volkan Sezer, Serhan Danış, Mehmet Karaca, Mehmet Abbak, Mustafa Gökhan Uzunbaş, Kayhan Eritmen, Mümin Imamoğlu, and Çağatay Karabat

Abstract In this chapter, we present data collection activities and preliminary research findings from the real-world database collected with "UYANIK," a passenger car instrumented with several sensors, CAN-Bus data logger, cameras, microphones, data acquisitions systems, computers, and support systems. Within the shared frameworks of Drive-Safe Consortium (Turkey) and the NEDO (Japan) International Collaborative Research on Driving Behavior Signal Processing, close to 16 TB of driver behavior, vehicular, and road data have been collected from more than 100 drivers on a 25 km route consisting of both city roads and The Trans-European Motorway (TEM) in Istanbul, Turkey. Challenge of collecting data in a metropolis with around 12 million people and famous with extremely limited infrastructure yet driving behavior defying all rules and regulations bordering madness could not be "painless." Both the experience gained and the preliminary results from still on-going studies using the database are very encouraging and give comfort.

Keywords Drive-Safe · NEDO · UYANIK · Data collection · Istanbul route · Brake pedal pressure sensor · Gas pedal pressure sensor · Laser ranger finder · EEG · CAN-Bus · IVMCTool · Driver modeling · Distraction · Fatigue · Safety · Abnormal driving · Reference driving · Online banking · Navigational dialog · Road-sign reading · Driver verification · Speech recognition · Multi-classifier · Facial feature tracking · 3D head tracking

3.1 Introduction

Throughout the world, driver error has been blamed as the primary cause for approximately 80% of traffic accidents. For instance, in 2005 there were more than 3,200 fatalities and 135,000 plus bodily injuries in over 570,000 traffic

H. Abut (✉)
San Diego State University (Emeritus), San Diego, CA, 92182, USA
e-mail: abut@anadolu.sdsu.edu

K. Takeda et al. (eds.), *In-Vehicle Corpus and Signal Processing for Driver Behavior*, DOI 10.1007/978-0-387-79582-9_3,

accidents in Turkey according to the Turkish Traffic Education and Research Directorate. Furthermore, it has been estimated that the long-term economic loss emanating from these accidents is over US $6 billion, which puts a rather significant burden on national budget. Albeit some variations, statistics from many other countries spanning the globe are very similar.

In 2004, Drive-Safe (DSC) Consortium, an academia and industry research partnership has been established in Turkey to create conditions for prudent driving on highways and roadways with the purposes of reducing accidents caused by driver behavior and to offer more amenities [1]. The objectives of this program consist of but are not restricted to

- DSP Technologies for driver identification based on multi-sensory behavioral and analog driving signals
- Driver modeling for personalization of driving environment
- Framework for automated detection of abnormal driving behavior from multiple sensors (audio, video, and driving signals)
- Development of technologies for "passive" assistance to mitigate abnormal driving and technologies for "active" assistance to reduce abnormal driving
- Ad hoc vehicle-to-vehicle communication for driving assistance

To achieve its objectives, collection of critical real-world data from multiple sensors (cameras, microphones, CAN-Bus, and other means) is required to build a coherent database on driver behavior, vehicle, and road conditions.

It is not difficult to guess characteristics in driving behavior differ from country to country according to their cultural and social backgrounds. Under the flagship of Nagoya University, an international alliance has been established in 2005 with several research groups in the United States, Japan, Turkey, Italy, and Singapore to share the worldwide driving, road, and vehicle-specific data obtained from 450 drivers (NEDO Alliance). Toward that end, three vehicles—"UYANIK" in Istanbul, "Nagoya Vehicle" in Japan (see Chapter 4), and "UT-Drive" in Dallas, Texas, USA (see Chapter 5)—have been equipped in cooperation with each other [2,3]. Data collection is underway in the three countries with potential of an ISO application in this area [4].

In this chapter, we will be focusing on the data collection effort with "UYANIK" and the preliminary findings by a number of members of the DSC in Turkey, who are also participating in the NEDO Alliance. The chapter is organized as follows. Section 3.2 discusses the instrumented vehicle, joint requirements of DSC/NEDO, sensors, and data acquisition systems. The selected data collection route, tasks, driver profile, and some challenges faced—both resolved and un-resolved "pains"—are addressed in Section 3.3. Signal samples are presented in Section 3.4. Next, research activities of Drive-Safe Consortium using collected data are highlighted in Section 3.5 as "Gains Are Coming." In continuing, preliminary findings from the student projects are mentioned in Section 3.6. Finally, Section 3.7 concludes the planned study with future work.

3.2 "Uyanik" and Sensors

Vehicle: As depicted in Fig. 3.1, "UYANIK,"—awake—is a Renault sedan donated to the consortium by OYAK-Renault of Turkey after many task-specific retrofitting done in the factory to serve the needs of this project. Modifications include special fortified front bumper, high-power battery, 1500 W DC–AC converter, CAN-Bus output socket, navigator seat and instrument bench, and re-wiring for power and signaling.

Sensors: The complete system layout for sensors, data acquisition systems, and wiring is shown in Fig. 3.2. There are two sets of cameras—one set for daylight and another for night vision—configured to give facial shots of the driver from both sides and a third one pointed to the road ahead as seen in Fig. 3.3. Four audio recordings are made by a lapel microphone or a headset, and two microphones on the rearview mirror and at the back of the head rest on the driver seat to capture the chamber noise and the conversation with the navigator. The human navigator is normally silent and handles the recording process except assistance when needed. Fourth microphone is the audio recording made from the microphone of the mobile phone and placed on the chest of the driver used for hands-free dialog.

An 180° laser range finder reads 181 x–y distances of the objects/vehicles in its range. Brake and gas pedal pressure readings from sensors connected to the pedals, several sensor outputs from the CAN-Bus socket of the vehicle, acceleration in xyz directions by an IMU device, location information from a GPS receiver are recorded at the sampling rate of the CAN-Bus. A 20-channel portable EEG subsystem is available for ground truth experiments.

Joint Requirements: In the data collection phase, the DSC/NEDO teams have come with the following desirable data set:

(1) **Three channels of uncompressed video:** They are captured (left and right view of the driver and the road ahead) at 30 frames per second. This corresponds to 140–200 GigaBytes (GB) of data per driver for frame-accurate audio–visual face tracking and recognition tasks. For Nedo Alliance applications, only two channels of MPEG-compressed video at 3.0 Mbits/s are recorded.

Fig. 3.1 "UYANIK" data collection vehicle

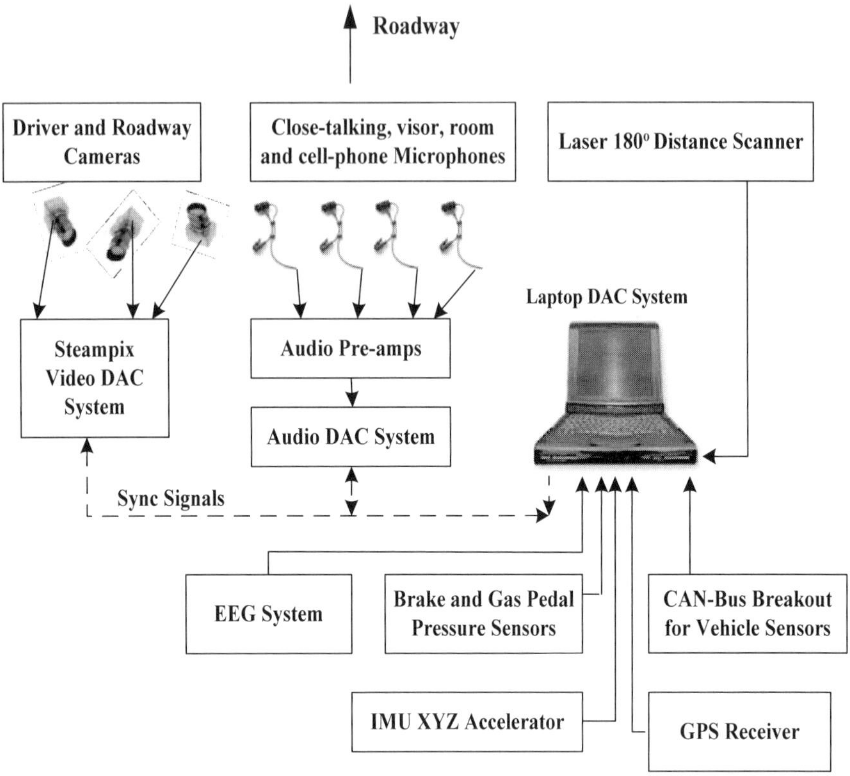

Fig. 3.2 System layout for sensors, data acquisition systems, and signal paths

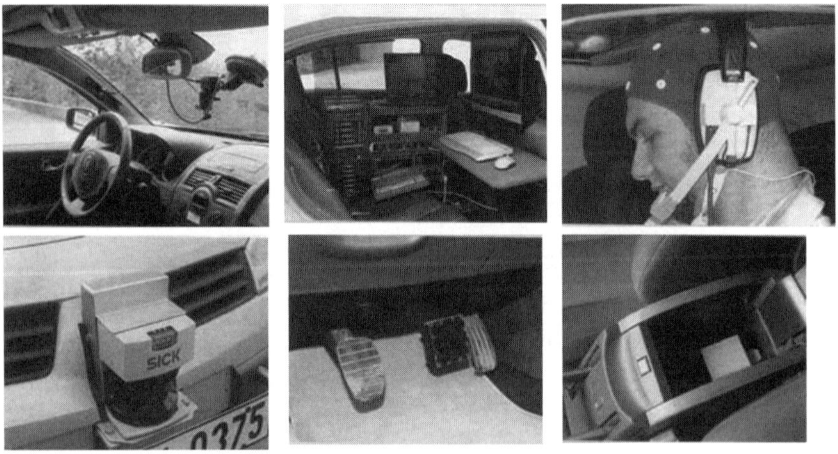

Fig. 3.3 Vehicle sensors: cameras, navigator area and instruments bench, laser range finder, 3D accelerator, brake pressure sensor, microphone, and EEG cap (from *top left clockwise*.)

(2) **Three audio recordings**: Lapel/headset, rearview mirror, chamber noise, and the dialog over the mobile phone. They are digitized at 16,000 samples per second with 16-bit resolution either in raw format or in wav format.

Vehicle speed, engine RPM, steering wheel angle, head distance, location, EEG (only for grown-truth experiments), brake pedal, and gas pedal status readings are to be recorded not more than 1.0 kHz sampling rate.

(3) **CAN-Bus readings:** Even though most manufacturers claim that their unit complies with the standards the formatting (data packaging) is proprietary. Brake pedal status (pressed/idle) and gas pedal engagement percent are to be recorded together with the speed and RPM at the rate permitted by the manufacturer. Renault Megane is designed to sample at either 10 or 32 Hz.
(4) **Pedal sensors:** Brake pedal and the gas pedal pressure sensor readings are digitized at the CAN-Bus sampling rate by an independent two-channel A/D. They are bundled with CAN-Bus signals and recorded in a laptop computer.
(5) **Site-specific sensors:** Different from Nagoya and Dallas data collection sites, UYANIK was equipped with a laser distance measuring device in the front bumper and also with an IMU XYZ Accelerator measuring sensor set-up. In addition, a number of drivers have used the vehicle with an EEG device to form the ground truth for experiments.

Data Acquisition Systems: Data are collected by three acquisition systems synchronized with each other.

Video Acquisition: Uncompressed digital recordings of video from three cameras with 30 frames per second and a frame size of 640 × 480 were achieved by a semi-custom StreamPix digital video recorder from NORPIX. Each video channel is recorded into a separate 750 GB HD with a separate firewire connection and the total HD budget per driver is about 150 GB. At the end of each week, the data are archived into HQ archiving tapes.

Audio Acquisition: Alesis ADAT HD24 Data Acquisition System is used for audio recordings. Four microphone channels and a sync signal between the two acquisition systems were sampled at 48 kHZ and 24 bits per sample. Later these are converted to 16 kHZ and 16 bits off-line in wav format.

Acquisition of CAN-Bus signals, laser range finder, GPS receiver, brake pedal sensor, and IMU XYZ Accelerator was realized over USB and RS232 PCMCIA ports of a notebook computer using a custom software developed by two of the authors listed above from the Mekar Labs and Autocom Center at the Technical University (ITU) of Istanbul—a DSC partner.

Data Acquired on a Notebook Computer: Engine speed, RPM, steering wheel angle, brake pedal status, percent gas pedal position, brake pedal pressure, gas pedal pressure, clutch status, rear gear engagement status, and individual tire speeds are recorded at 32 samples per second. It is worth noting that no more than 10 Hz sampling rate was possible at the beginning of experiments and the data sets for some drivers need to be re-converted to 32 Hz for uniformity. In addition, the laser range finder readings (1 per second), IMU readings, and location information are recorded by this notebook. Undependable behavior of the GPS receiver was overcome by a second location tracking system donated by Satko, Inc. of Istanbul, a DSC sponsor. This alternate location information is tracked and recorded at the base control center located at the ITU campus.

3.3 Tasks and Pains

Route: Data collection starts and ends at the OTAM Research Center in the ITU Campus in Ayazağa, where UYANIK was housed. The navigator and the server for data storage are also located there. The data collection route is little over 25 km. It consists of a short ride inside the campus, followed by two 1.5 km very busy city thoroughfare sections, where a major construction is taking place to build a metro station. TEM Motorway toward the airport is the next segment. The route exits the first exit and makes a U-turn and travels toward the FSM Bridge. Highway driving ends at the Etiler exit and the rest of the route goes through city streets in Etiler, Akatlar, Levent, 4.Levent, Ayazağa, and back to OTAM at ITU campus via the ever-busy Büyükdere Caddesi (Fig. 3.4).

Data collection in this route has been a major challenge in a city of 12 million, famous with extremely limited infrastructure, drivers defying all rules and regulations, and the complete lack of courtesy to other vehicles and pedestrians around. Hence, the driving experience can be best portrayed as an art bordering madness.

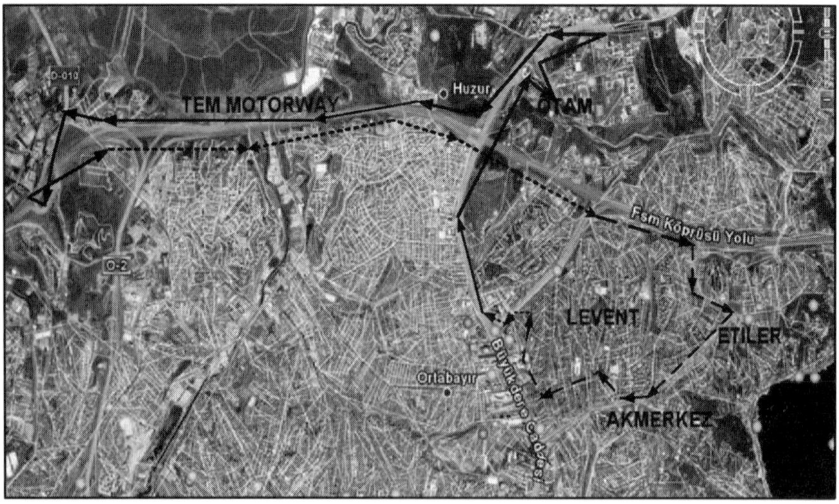

Fig. 3.4 DSC/NEDO route for Istanbul data collection. (Legend: *Solid lines*: free driving and radio tuning; *dotted lines*: on-line banking using ASR; *dashed lines*: Road/building signs, navigational dialog with human operator, and license-plate reading

Data Collection Tasks: As marked on the route map, there are four primary tasks each driver goes through:

(1) *Reference driving*: Here the driver gets used to the vehicle, the route, and the tasks. Most drivers turn on the radio and they tune to their favorite station. It was planned to have an ASR-based music query in this segment. A homegrown package was experimented but the results were not satisfactory.
(2) *Query dialog*: In this segment, each driver performs on-line banking using the cell phone mounted on the dashboard, which is programmed for speed dialing. He/she queries market prices of several national stocks using the on-line ASR-based service of a national bank. Here is a synopsis of the dialog:

ASR:	*Please tell the full name of the stock after beep.*
Driver:	*Arcelik*
ASR:	*6 YTL & 58*. (If successful, female voice)
ASR:	*I did not understand please repeat after beep*. (If unsuccessful, male voice)

(3) *Signboard reading and navigational dialog*: The driver announces the road and other signs posted on boards and buildings. At Etiler exit on TEM, base station is dialed and the next phase of the route is verified. Sign reading continues for about 2 km together with the license plates of the vehicles around. Both audio signals from the input and the speaker of the cell phone are recorded.
(4) *Pure navigational dialog*: After completing the very busy segment of the road in Etiler, the driver frequently contacts the navigator and conducts a human-to-human dialog.

Upon completion of the final segment of the route on Büyükdere Caddesi (again free driving), the experiment ends in front of the OTAM where the driver completes a couple of forms and a questionnaire.

Driver Profile: In Istanbul data collection effort, 108 drivers (100 were required) have driven the vehicle in the 25 km route, 19 of them were female and the remaining 89 male. The age range for female drivers was 21–48, and the corresponding male range was 22–61. Driver population was mostly pulled from the academic partners of the DSC together with their family and friends. However, due to equipment malfunction and a major disk crash affected data from several drivers. This brings the usable driver size to 101 in total.

Challenges Faced: As it was mentioned earlier, data collection on live traffic with human subjects is a major challenge by itself, which should not be seen as a natural extension of experiments conducted with simulators in a controlled lab environment. They are affected by equipment failures, technical difficulties, climate and weather, traffic density and type, driving local culture, effectiveness of law enforcement, and probably the most critical ones are the driver's physical and mental conditions and his/her driving behavior. In this undertaking, there were challenges in each area and many of them came all at once. Some examples with their current status/solution in parenthesis are

- Digital recording of uncompressed video at 30 frames/s with 640 × 480 resolution. (Solution: three-channel StreamPix system.)
- Synchronization of three different data logging systems. (Solution: common system clock and sharing a sync signal.)
- Data store archives approximately 16 TB of data from 108 drivers and their backup into tapes. (Solution: employment of a full-time engineer and part-time students to archive data into HQ digital back tapes on an HP backup subsystem. Each experiment of 40–50 minutes has required 4–5 hours to access, align, and write into a tape.)
- Off-line compression of data into MPEG4 format and alignment of various sensor readings. (Solution: custom-designed software package for aligning, compressing, and browsing/re-recording.)
- CAN-Bus data readings. (Solution: special software package.)
- CAN-Bus data cannot be read at a programmable rate, it fluctuates around, either 10 or 32 Hz. Earlier experiments were done at about 10 Hz and the final data sets were collected at 32 Hz. (Solution: rate for each run is recorded and re-sampling is done subsequently by users of the database.)
- Unacceptably noisy signal from brake/gas sensor pedal pressure. (Solution: semi-custom-designed A/D for these signals.)
- More reliable location information (unresolved).
- ASR for Music Search, i.e., query (under study).
- Complaints from drivers using pedal with sensors mounted. (Solution: this is mitigated by opening the gap between pedals at the Renault manufacturing plant and advising drivers to use light-weight shoes.)
- Complaints from drivers in "multi-tasking" and losing attentiveness. (Solution: do on-line banking on the curb side or earlier inside ITU campus if they feel losing control.)
- Wrong turns from unfamiliarity with the route, which results in significant time loss and the stored data size. (Solution: emergency assistance by the navigator and frequent assistance from the control center at OTAM via cell phone.)

3.4 Signal Samples

A screen shot from a male driver is shown in Fig. 3.5. Here, video feeds from two cameras mounted in the vehicle pointing at the driver and a third one toward the road showing a residential neighborhood along the route are displayed together with four speech waveforms down-sampled to 16 kbits/s and 16-bit resolution (recorded in wav format). Driving signals fed from the CAN-Bus are also displayed. Video feeds could be either uncompressed or compressed with MPEG4 at 3 MB/s avi format using DivX codec.

Fig. 3.5 Browser synopsis of the multi-channel multimedia display for Driver IF1014. Three video feeds and four audio waveforms together with steering wheel angle, vehicle speed, and engine RPM dials are included

As can be seen from Fig. 3.5, steering wheel angle in degrees, vehicle speed in km/h, and the engine RPM are illustrated. There are four visible dots on the needle of speed dial, which represent the actual tire speeds. Normally, they move together. However, they register different values during skids and slides in rainy weather or inclined surface, and sudden brakes.

Brake and gas pedal pressures are displayed in Fig. 3.6. In addition, the vehicle speed and the steering wheel angle (sampled at 34 Hz) are shown in Fig. 3.7 for an interval of 640 seconds from a test run. Vehicle-related signals recorded from CAN-Bus data are also synchronized with video but the re-sampled versions are not recorded, i.e., originals are kept. There are few other readings from the CAN-Bus reporting the status of clutch, rear gear, and brake pedal.

One hundred and eighty one distance measurements between UYANIK and other vehicles/objects around are measured by a laser ranger finder at the rate of 1 Hz and Fig. 3.8 shows both the plot and the actual photo at that instant, which is explicitly marked.

In addition, steering wheel angle velocity, yaw rate, clutch status, and rear gear status readings are recorded from the CAN-Bus.[1] IMU readings showing

[1] These readings are information-rich for projects carried out with active/passive vehicle control and avoidance systems.

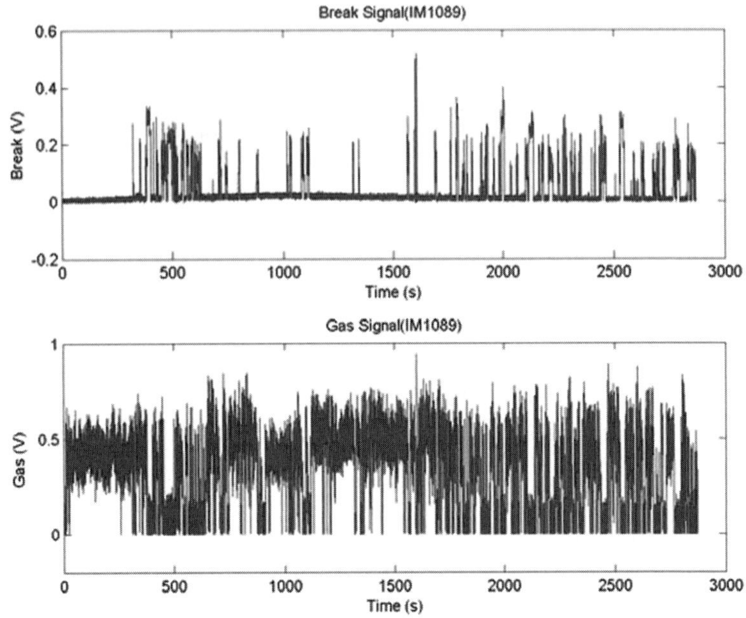

Fig. 3.6 Brake and gas pedal pressure readings

the *xyz* directional accelerations and the location information from GPS receiver are recorded into separate files. A number of application-specific plots are illustrated in several related works briefly discussed in Section 3.6.

Finally, we have recorded in a number of experiments 16-channel EEG signals to act as ground-truth for fatigue and attention/inattention determination. For the driver shown in Fig. 3.9, plots from eight channels—specifically, C3, C4, F3, F4, O1, O2, A1, and A2, which are important for fatigue prediction—are shown in Fig. 3.9. On the other hand, EOG channels are recorded for the estimation and the rejection of eye blinks. ECG and EMG are also used for both artifact rejection and fatigue estimation. Alpha signal power and signal power ratio between right and left hemispheres are used for fatigue prediction from EEG measurements after rejection of ECG, EMG, and EOG artifacts, the importance of these are currently under study. For fatigue and distraction understanding, long-run simulator experiments (3–4 hours long or more) with/without EEG are in progress.

3.5 Gains are Coming: Part 1

Using the database generated during the data collection process, the preliminary findings from three on-going projects carried by scholars and their students at Sabancı University in Istanbul, Turkey, will be briefly discussed in this and the following sections.

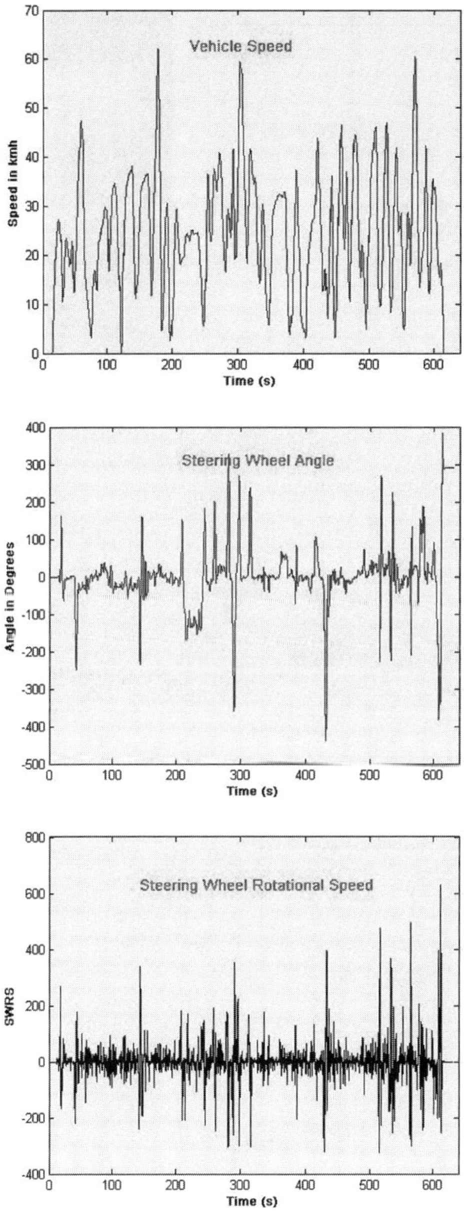

Fig. 3.7 Vehicle speed, engine, steering wheel, angle, and steering wheel rotational speed plots are for the first 380–560 seconds of one test

3.5.1 Audio–Visual Speech Recognition in Vehicular Noise Using a Multi-classifier Approach by H. Karabalkan and H. Erdoğan

Speech recognition accuracy can be increased and noise robustness can be improved by taking advantage of the visual speech information acquired from

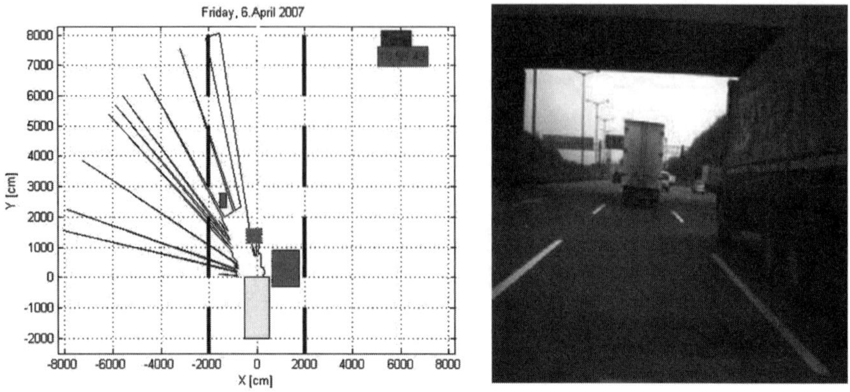

Fig. 3.8 Laser range finder readings and the view of Driver IF1015. Truck on the right is between −200 to +800 cm relative to UYANIK (overlapping by 2 m and another 8 m ahead and the white truck in front is 12 m ahead)

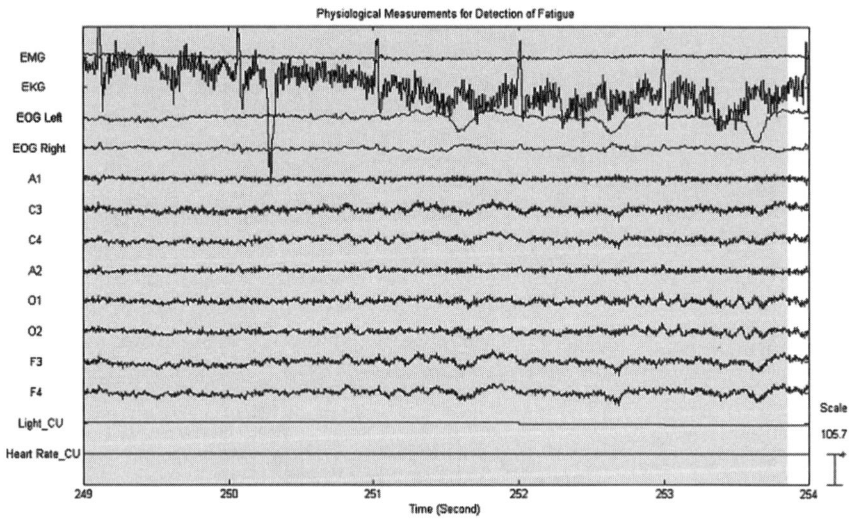

Fig. 3.9 Eight channels of the EEG recordings for the driver in Fig. 3.3

the lip region. To combine audio and visual information sources, efficient information fusion techniques are required. In this paper, we propose a novel SVM–HMM tandem hybrid feature extraction and combination method for an audio–visual speech recognition system. From each stream, multiple one-versus-rest support vector machine (SVM) binary classifiers are trained where each word is considered as a class in a limited-Vocabulary speech recognition scenario.

The outputs of the binary classifiers are treated as a vector of features to be combined with the vector from the other stream and new combining binary classifiers are built. The outputs of the classifiers are used as observed features in hidden Markov models (HMM) representing words.

The complete process can be considered as a nonlinear feature dimension reduction system which extracts highly discriminatory features from limited amounts of training data. To simulate the performance of the system in a real-world environment, we add vehicular noise at different SNRs to speech data and perform extensive experiments.

In the approach reported here, 12-dimensional cepstral coefficients and the energy in the window are extracted to obtain static MFCC coefficients. Then, the feature vector is extended to a 39-dimensional vector by taking the first and second time differences of static coefficients. The audio feature vector is passed through a multiple binary classifier structure, as shown in Fig. 3.10, to obtain an 11-dimensional audio feature vector which is more discriminative and noise tolerant. The dimension is 11 since there are 10 different words uttered in the database and an additional class is considered for silence.

To extract visual features, principal component analysis (PCA) is applied to the lip region which returns the eigenlips. Only the 30 most significant eigenlips are considered to come up with 30-dimensional visual feature vector. This reduces dimensionality as well as speaker independence. The 30-dimensional visual feature is also passed through a multiple binary classifier structure as in the case of audio features, and finally an 11-dimensional visual feature vector is obtained.

Multiple paralleled binary classifier structure is used for combining audio and visual information, too. The audio feature vector and the visual feature vector, which are both 11-dimensional, are concatenated to form a 22-dimensional audio–visual feature vector and this feature vector is passed through the multiple binary classifier structure. The resulting 11-dimensional audio–visual

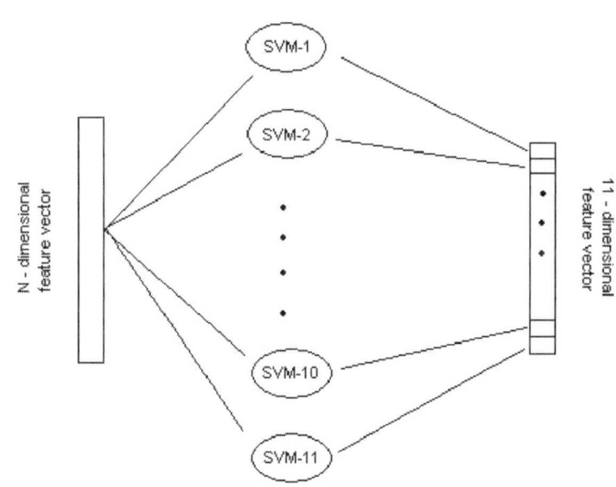

Fig. 3.10 Multiple parallel binary classifier structure

feature vector is the observation for HMM. The proposed fusion technique is planned to be compared with linear discriminant analysis (LDA). The experiments for SVM-HMM tandem hybrid approach are still on-going but the method promises good accuracy rates throughout different SNR values [5].

3.5.2 Graphical Model-Based Facial Feature Point Tracking in a Vehicle Environment by S. Coş ar

Facial feature point tracking is an important step in problems such as video-based facial expression analysis, human–computer interaction (HCI), and fatigue detection. Generally, such analysis systems consist of three components: feature detection, feature tracking, and expression recognition. Feature detection involves detecting some distinguishable points that can define the movement of facial components. This may involve detection of eyes, eye brows, mouth, or feature points of these components. Next is the tracking part which consists of tracking the detected feature points. Finally, according to tracking results of these feature points, the recognition component outputs results such as happy, sad, or tired.

For feature point tracking, roughly there are two classes of methods in literature: general purpose approaches and face-specific approaches. Generally, feature point tracking is done by using a temporal model that is based on pixel values. Consequently, these methods are sensitive to illumination and pose changes, and ignore the spatial relationships between feature points. This affects the tracking performance adversely, causes drifts and physically unreasonable results when the data are noisy or uncertain due to occlusions.

In this work, feature point tracking is performed in a statistical framework that incorporates not only temporal information about feature point movements but also information about the spatial relationships between such points. This framework is based on graphical models that have recently been used in many computer vision problems. The model is based on a parametric model in which the probability densities involved are Gaussian. The parametric nature of the models makes the method computationally efficient. Spatial connections between points allow the tracking to continue reasonably well by exploiting the information from neighboring points, even if a point disappears from the scene or cannot be observed. Feature values from video sequences are based on Gabor filters. Filters are used in a way to detect the edge information in the image, to be sensitive to different poses, orientations, and feature sizes. Based on this model, an algorithm that achieves feature point tracking through a video observation sequence is implemented. The current method is applied on 2D gray scale real video sequences taken in the vehicle environment, UYANIK, and the superiority of this approach over existing techniques is demonstrated in Figs. 3.11 and 3.12.

Fig. 3.11 Ideal environment tracking results from earlier methods (*left*) and the proposed method (*right*) sample shots from a laboratory setting

3.5.3 3D Head Tracking Using Normal Flow Constraints in a Vehicle Environment by B. Akan

Head tracking is a key component in applications such as human–computer–interaction, person monitoring, driver monitoring, video conferencing, and

Fig. 3.12 Driver head tracking based on an existing method (*left*) and the proposed method (*right*). Video captured from UYANIK

object-based compression [6]. The motion of the head of a driver head can tell a lot about his/her mental state, e.g., whether he/she is drowsy, alert, aggressive, comfortable, tense, or distracted. This chapter reviews an optical flow-based method to track the head pose, both orientation and position, of a person and presents results from real-world data recorded in a car environment.

Driver behavior modeling and fatigue detection is an important feature in developing new driver assistance systems and smart cars. These intelligent vehicles are intended to be able to warn or activate other safety measures when hazardous situations have been detected such as fatigued or drunk driver, so that a system can be developed to actively control the driver before he/she becomes too drowsy, tired, or distracted. The pose of the head can reveal numerous clues about alertness, drowsiness, or whether the driver is comfortable or not. Furthermore knowing the pose of the head will provide a basis for robust facial feature extraction and feature point tracking.

In this, study, we propose a method for tracking the driver's head using normal flow constraint (NFC) which is an extension of the original optical flow algorithm. Optical flow is the 2D vector field which is the projection of the 3D motion onto an image plane. It is often required to use complex 3D models or nonlinear estimation techniques to recover the 3D motion when depth information is not available. However, when such observations are available from devices such as laser range finders (laser scanners) or stereo cameras, 3D rigid body motion can be estimated using linear estimation techniques [7].

We have tested the algorithm in a real car environment. A bumblebee stereo camera system has been used for data acquisition. The camera hardware analyzes the stereo images and establishes correspondence between pixels in each image. Based on the camera's geometry and the correspondences between pixels in the images, it is possible to determine the distance to points in the scene. Without any special optimizations the tracker can update pose estimations based on 2,000–3,000 pixels per frame at a rate of 60 Hz on a Celeron 1.5 GHz laptop.

Performance of the tracker has been tested using the data collected from "UYANIK." Several sequences of length 500 frames or roughly 30s of video with both intensity and disparity images have been recorded. The sequences involve all natural head movements: throughout the video the driver rotates his head checking out left, right, and rear mirrors of the car and looks down at the gear. Some outputs from the tracking algorithm can be seen in Fig. 3.13.

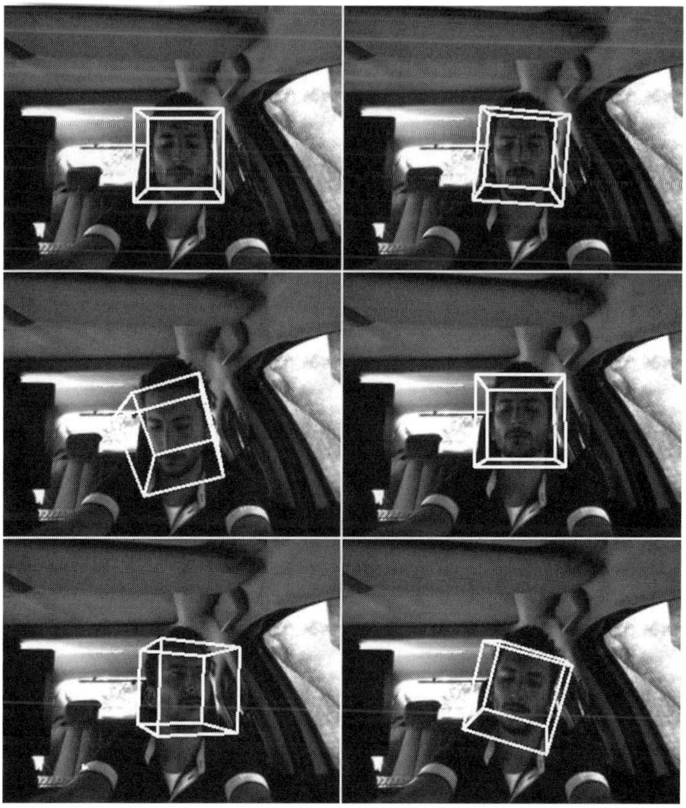

Fig. 3.13 Results of the driver head tracker at frames 0, 60, 110, 150, 230, 400

3.6 Gains are Coming: Part 2

Along with the three applications in the previous section, two additional projects are under study by several students at Sabancı University, Istanbul, Turkey. Because, these studies are still in progress, we will briefly identify these problems and present preliminary findings.

3.6.1 Pedal Engagement Behavior of Drivers by M. Karaca, M. Abbak, and M.G. Uzunbaş

In Fig. 3.14, we illustrate a sample of brake pedal and the gas pedal engagement status for a driver and the correlation between these two waveforms, we obtain zero-crossings per minute (zero-crossing rate) indicating the transitions in the

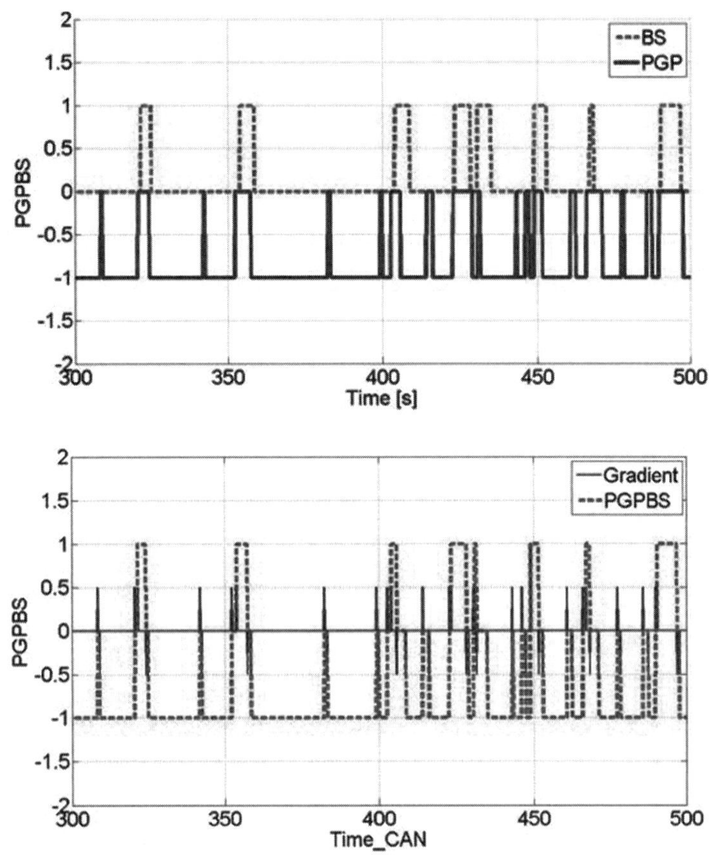

Fig. 3.14 Brake and gas pedal status plots in a 200-s interval

pedal engagement from brake-to-gas and vice versa (pedal shift). Also, the head distance between the test vehicle and other vehicles ahead is depicted in Fig. 3.8.[2]

This multi-sensory information is explored for understanding the gender-specific behavior of drivers to the traffic ahead by obtaining the statistics of intra-/inter-gender zero-crossings per minute rates for several male and female drivers in the database. Interesting patterns have been observed but not ready to interpret them until more extensive studies are done. However, one local folk theorem is proven.

The average inter-gender pedal shifts per minute at various speeds are shown in Fig. 3.15. It is clearly observable that the female drivers in Turkey are driving more smoothly across all speeds when compared to their male counterparts. Furthermore, male drivers are very impatient and make frequent brake/gas shifts at a wide range of speeds of 40–80 km/h. Should the male drivers in Turkey leave the driving to ladies?

3.6.2 Speaker Verification and Fingerprint Recognition by K. Eritmen, M. Imamoğlu, and Ç. Karabat

In this application, drivers use their fingerprint and speak their names to start the vehicle. The purpose of this application is twofold: (a) access/deny to the vehicle with one physical and one behavioral signature—fingerprint and speech—for improved performance and (b) still access/deny in the case if only one of the sensory modes is available.

In biometrics, signal processing, and forensic communities, performance of biometric verification systems is measured by using receiver operating characteristics (ROC) curve which is the plot of false accept rate (FAR) versus false reject rate (FRR) for changing threshold values [5].

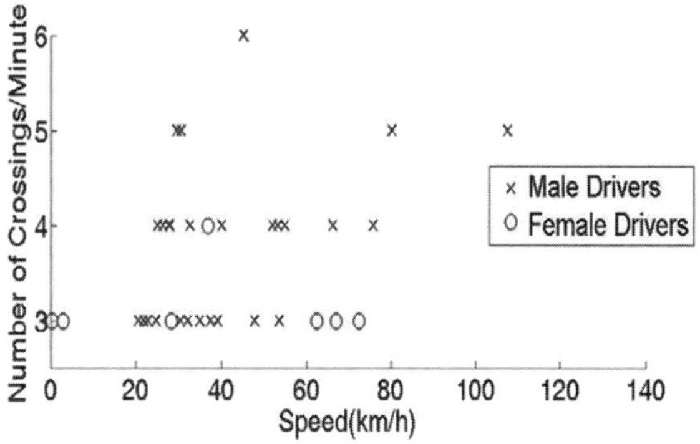

Fig. 3.15 Gas and brake pedal zero-crossing rates for female and male drivers

[2] Vehicle identification at a given time is done manually by studying the picture and the coordinates of the distances to the objects recorded by the laser scanner simultaneously.

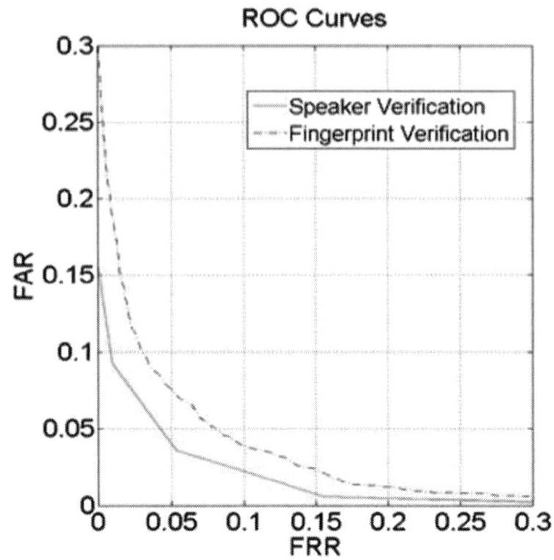

Fig. 3.16 Individual ROC curves for fingerprint and speaker verification before score-level fusion

In Fig. 3.16, we show the false acceptance (FAR) versus false reject (FRR) rates obtained for several drivers. In these experiments, we have used the familiar GMM-based speaker models obtained from MFCCs. Individual equal error rates are approximately 4–7% for these two signatures, which are very similar to results reported in the literature.

At the time of this writing, reliable and meaningful score-level classifier experiments were performed to fuse the results with the anticipation of improved performance as it was reported in many data fusion application classification and in person/object identification problems including the works by some of the authors [4,5].

3.7 Conclusions and Future Work

In this chapter, we report the progress on real-world data collection with "UYANIK" in Istanbul (Turkey) as part of an international collaboration research with Nagoya University (Japan) and the University of Texas in Dallas (USA) and in partnership with the Drive-Safe Consortium in Turkey. A total of 101 drivers participated in the experiment to date resulting at a data storage size of more than 16 TB with very rich information on driver behavior, the vehicle performance, and the road and traffic conditions on a 25.6 km route in Istanbul.

We will complement the road data with experiments carried on a recently installed full-fledged driving simulator with research focus on driver modeling, fatigue, and distraction detection. At the same time, we would transcribe and analyze cognitive task-loaded and free-driving segments to better understand the impact of certain tasks on drivers and to make technology recommendations for safer driving.

Acknowledgments This work is partially supported by the State Planning Organization of Turkey (DPT) under the umbrella initiative called "Drive-Safe Consortium," the NEDO collaborative grant titled "International Research Coordination of Driving Behavior Signal Processing Based on Large Scale Real World Database" from Japan, and the European Commission under grant FP6-2004-ACC-SSA-2 (SPICE).

The authors would like to acknowledge Mr. Hakan Tandoğdu, his team at Renault Manufacturing Plant in Bursa for retrofitting UYANIK, and his management for donating the vehicle. Turkcell, OPET, and Satko of Turkey have been sponsoring the data collection effort with goods and services.

We kindly appreciate the contributions and advice from Dr. Levent Guvenc and his students at ITU, Dr. Ali G. Göktan and Yunus Canlı at the OTAM Center of ITU in Turkey, and Dr. Engin Erzin of Koç University in Istanbul, and Dr. John H.L. Hansen and his team in Dallas. Without Dr. Kazuya Takeda of Nagoya University in Japan, the undertaking of this magnitude could not have happened. We are indebted to him and his colleagues in Nagoya.

References

1. A. Erçil, H. Abut, E. Erzin, E. Göçmençelebi, A. Göktan, L. Güvenç, E. Özatay, H. Tandoğdu, "The DriveSafe Project," Proceedings, Autocom Workshop on Preventive and Active Safety Systems for Road Vehicles, September, 19–20, 2005, Istanbul, Turkey.
2. C. Miyajima, T. Kusakawa, T. Nishino, N. Kitaoka, K. Itou, K. Takeda, "On-Going Data Collection of Driving Behavior Signals," Chapter 4 in this book.
3. P. Angkititrakul, J.H.L. Hansen, "UTDrive: The Smart Vehicle Project," Chapter 5 in this book.
4. H. Abut, J.H.L. Hansen, K. Takeda (Editors), *Advances for In-Vehicle and Mobile Systems: Challenges for International Standards*, Springer Science-Business, May 2007, ISBN: 0-387 33503-X.
5. E. Erzin, Y. Yemez, A. M. Tekalp, A. Erçil, H. Erdoğan, H. Abut, "Multimodal Person Recognition for Human–Vehicle Interaction," *IEEE Multi-Media Transactions*, April–June 2006, pp. 18–31.
6. L.P. Morency, A. Rahimi, N. Checka, T. Darrell, "Fast Stereo-Based Head Tracking for Interactive Environments," Proceedings of Conference on Automatic Face and Gesture Recognition, 2002.
7. M. Harville, A. Rahimi, T. Darrell, G.G. Gordon, J. Woodfill, "3D Pose Tracking with Linear Depth and Brightness Constraints," International Conference on Computer Vision, pp. 206–213, 1999.

4
On-Going Data Collection of Driving Behavior Signals

Chiyomi Miyajima, Takashi Kusakawa, Takanori Nishino, Norihide Kitaoka, Katsunobu Itou, and Kazuya Takeda

Abstract We are developing a large-scale real-world driving database of more than 200 drivers using a data collection vehicle equipped with various sensors for the synchronous recording of multimedia data including speech, video, driving behavior, and physiological signals. Driver's speech and videos are captured with multi-channel microphones and cameras. Gas and brake pedal pressures, steering angles, vehicle velocities, and following distances are measured using pressure sensors, a potentiometer, a velocity pulse counter, and distance sensors, respectively. Physiological sensors are mounted to measure driver's heart rate, skin conductance, and emotion-based sweating. The multimedia data is collected under four different task conditions while driving on urban roads and an expressway. Data collection is currently underway and to date 151 drivers have participated in the experiment. Data collection is being conducted in international collaboration with the United States and Europe. This chapter reports on our on-going driving data collection in Japan.

Keywords Driving behavior · Multimedia database · Synchronous recording · Data collection vehicle · Real-world driving · Urban road · Expressway · Distraction · Physiological signal · Driving signal · Speech recognition · Spoken dialog · Speech interface

4.1 Introduction

The automobile industry is becoming one of the most important industries for future economic development in the world. With the increasing emphasis on driving safety, comfort, and convenience, advanced driver assistance systems including adaptive cruise control, lane-keeping assist systems, and car navigation systems with speech interfaces have been developed over the last few

C. Miyajima (✉)
Nagoya University, Furo-cho, Chikusa-ku, Nagoya, 464-8603, Japan
e-mail: miyajima@is.nagoya-u.ac.jp

K. Takeda et al. (eds.), *In-Vehicle Corpus and Signal Processing for Driver Behavior*, DOI 10.1007/978-0-387-79582-9_4,

decades. Future research directions will also focus on developing intelligent technologies for enhancing interaction between humans and vehicles.

For such research purposes, we are constructing a large-scale real-world multimedia driving database. We designed a data collection vehicle equipped with various sensors for the synchronous recording of speech, video, driving behavior, and physiological signals. Driver speech is recorded with 12 microphones distributed throughout the vehicle. Face images and a view of the road ahead are captured with three CCD cameras. Driving behavior signals including gas and brake pedal pressures, steering angles, vehicle velocities, and following distances are recorded. Physiological sensors are mounted to measure driver's heart rate, skin conductance, and emotion-based sweating on the palm and sole of the foot for detecting stress.

Multimodal data are collected while driving on city roads and an expressway under four different task conditions: reading words on signs and billboards while driving, being guided to an unfamiliar place by a human navigator on a cell phone with a hands-free device, reading random four-character alphanumeric strings by repeating after hearing, and interacting with a spoken dialog system to retrieve and play music.

The characteristics of driving behavior differ from country to country based on their cultural and social backgrounds. We are collaborating internationally with research groups in the United States and Europe to share the worldwide driving data of various drivers [1, 2]. Similar kinds of data collection vehicles have also been developed in the United States and Europe. Data collection is currently underway in the three areas of the world.

The multimodal database will be published for research purposes such as noise robust speech recognition in car environments [3, 4], detection of driver's stress while driving [5–7], and the prediction of driving behaviors to improve intelligent transportation systems. This chapter reports on our on-going driving data collection in Japan.

4.2 Design of Data Collection Vehicle

A data collection vehicle was designed for synchronous recording of driving data. The configuration of the recording system is described as follows.

4.2.1 Vehicle

TOYOTA Hybrid ESTIMA with 2,360 cc displacement (Fig. 4.1) was used for data recording. Various sensors and synchronous recording systems are mounted on the instrumental vehicle (Fig. 4.2). The design of the recording system is arranged as shown in Fig. 4.3.

Fig. 4.1 Vehicle used for data collection (TOYOTA Hybrid ESTIMA)

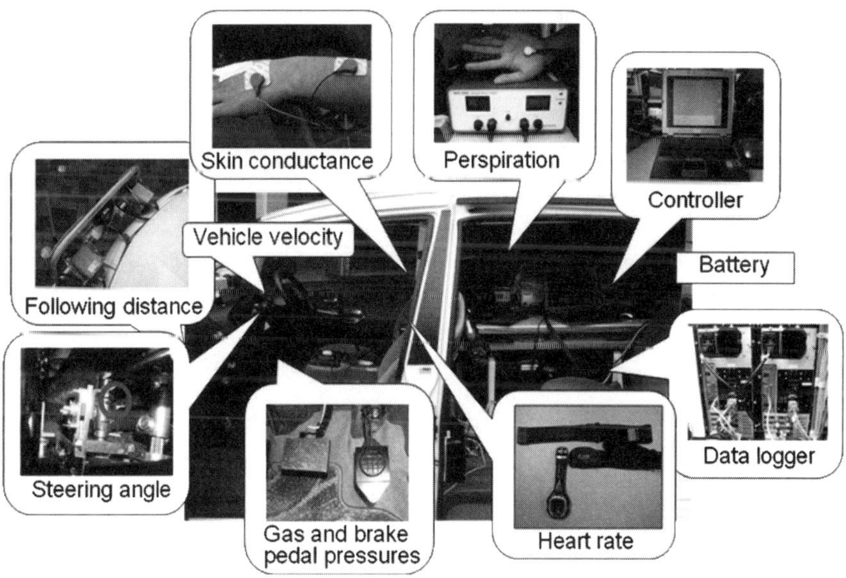

Fig. 4.2 Sensors mounted on the instrumented vehicle

4.2.2 Microphones

Eleven omnidirectional condenser microphones (SONY ECM-77B) and a close-talking headset microphone are mounted on the vehicle to record driver's speech. The recorded speech is amplified through YAMAHA amplifiers. The positions of the microphones are shown in Fig. 4.4.

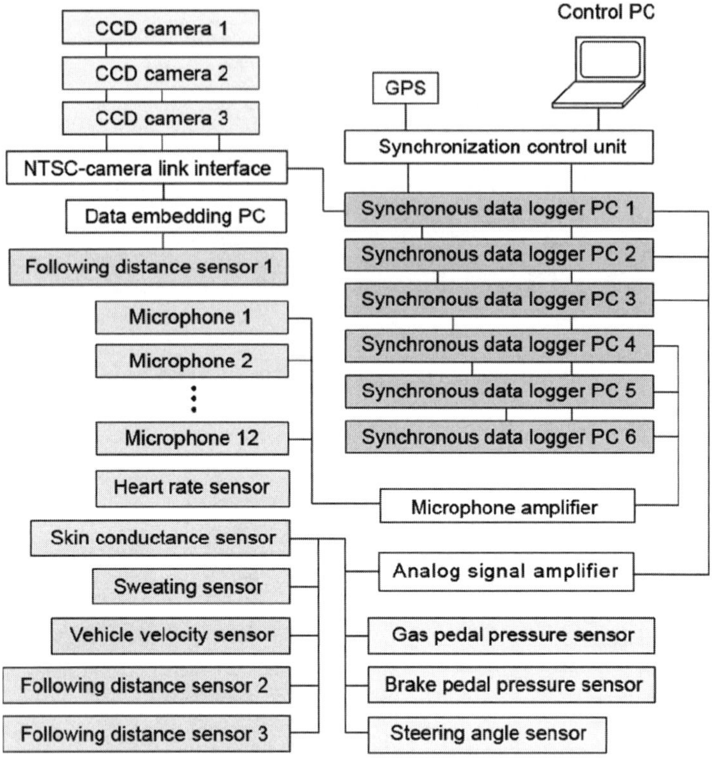

Fig. 4.3 Block diagram of recording system

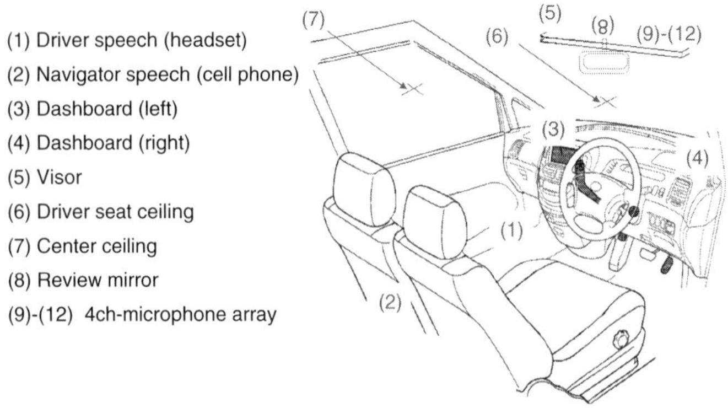

Fig. 4.4 Microphone positions

4.2.3 Video Cameras

Driver's face images from the right and left front and the view of the road ahead are captured with three CCD cameras (SONY DXC-200A) at 29.432 fps. The positions of the cameras are shown in Fig. 4.5. Figure 4.6 shows examples of video frames for cameras #1, #2, and #3.

4.2.4 Sensors for Driving Operation Signals

Driving signals of steering angles and brake and gas pedal pressures are recorded. A potentiometer (COPAL M-22E10-050-50 K) is used to measure steering angles, and pressure sensors (LPR-A-03KNS1 and LPR-R-05KNS1, respectively) are mounted on the gas and brake pedals.

4.2.5 Vehicle Status Sensors

The vehicle velocity is measured based on the output of the JIS5601 pulse generator. Distance per 100 ms is obtained by multiplying pulse intervals and tire circumference. Digital signals are converted to analog signals by a D/A converter.

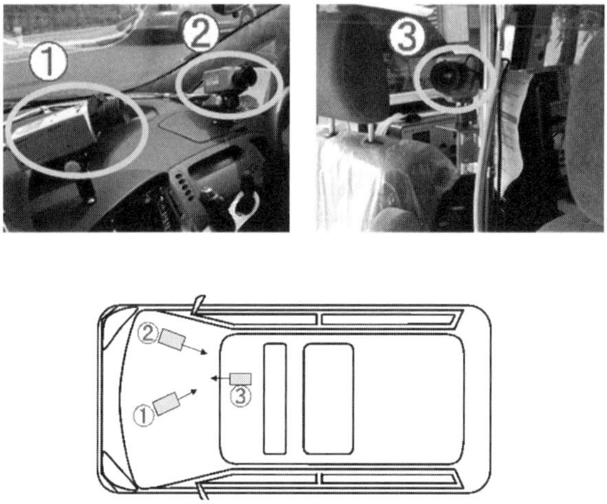

Fig. 4.5 Video camera positions

Fig. 4.6 Samples of video frames for three cameras

4.2.6 Vehicle Position Sensors

The following distance from a lead vehicle is measured by distance sensors. Two kinds of distance sensors (SICK DMT-51111 and MITSUBISHI MR3685) are mounted in front of the vehicle for measuring short and long ranges, respectively.

4.2.7 Physiological Sensors

Physiological signals are correlated to driver's stress [5]. Physiological sensors are mounted to measure driver's heart rate, emotion-based sweating on the palm and sole of the foot, and skin conductance.

Driver's heart rate is measured using a chest belt sensor (POLAR S810i), the amount of sweating is measured through a perspiration meter (SKINOS SKD-2000), and skin conductance is measured with an electrodermal meter (SKINOS SK-SPA).

4.2.8 Synchronous Recording System

For synchronous recording of the above signals, a multi-channel synchronous recording system (CORINS, MVR-303) is used. MVR-303 has a synchronous control unit and a system control PC and can record multi-channel synchronous videos and analog signals. Each PC node can store 240 GB video data of 1.4 million pixels and 29.432 fps that corresponds to 90-min videos.

4.3 Data Collection

To develop a technique for quantifying the stress level of drivers, driving data are recorded under various conditions with four different tasks.

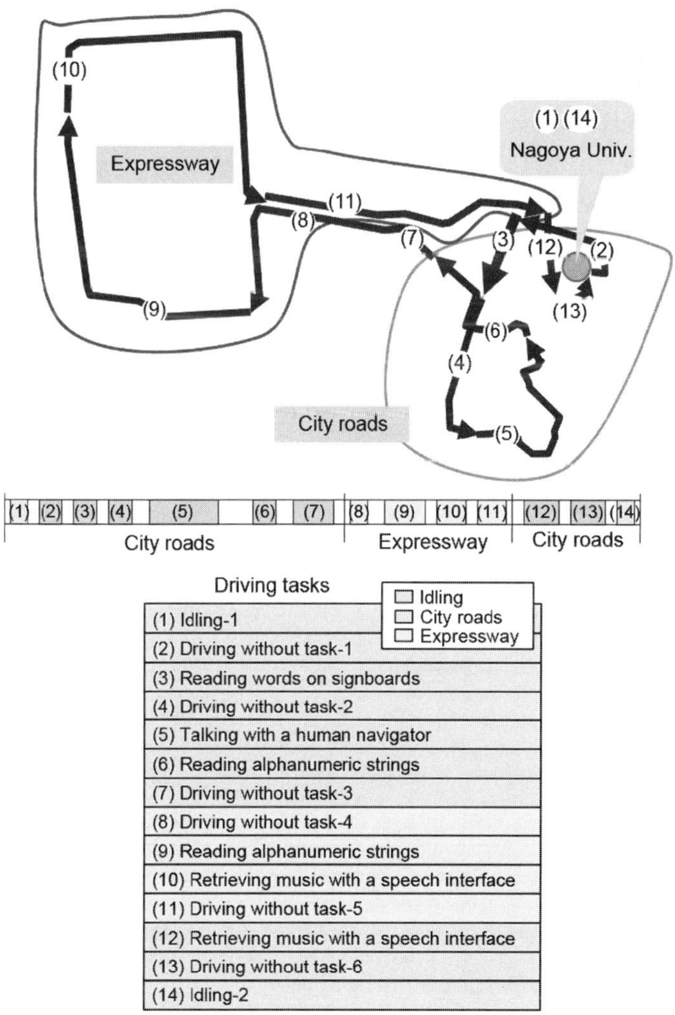

Fig. 4.7 Nagoya route used in data collection

4.3.1 Tasks

The details of the tasks are described as follows with examples of spoken sentences:

(1) *Signboard reading task*: Drivers read aloud words on signboards such as names of shops and restaurants seen from the driver seat while driving, e.g., "7–11" and "Denny's."
(2) *Navigation dialog task*: Drivers are guided to an unfamiliar place by a navigator on a cell phone with a hands-free headset. Drivers do not have maps, and only the navigator knows the route to the destination. The following is an example of a spoken dialog:
    Navigator: You should see a restaurant on your left.
    Driver: Yes, I see Kobeya.
    Navigator: Well, yeah, umm, you are at the Hibarigaoka intersection. Turn left at the intersection.
    Driver: O.K. I'll turn left.
(3) *Alphanumeric reading task*: Drivers repeat random four-letter strings consisting of alphabet a–z and digits 0–9, e.g., "UKZC," "IHD3," and "BJB8." The instruction of the four-letter strings is heard through an earphone.
(4) *Music retrieval task*: Drivers retrieve and play music from 635 titles of 248 artists by a spoken dialog interface. Music can be retrieved by artist name or song title, e.g., "Beatles" or "Yesterday."

Each driver starts from Nagoya University and returns after about 70 min of driving. The route map for data collection is shown in Fig. 4.7. Driving data are

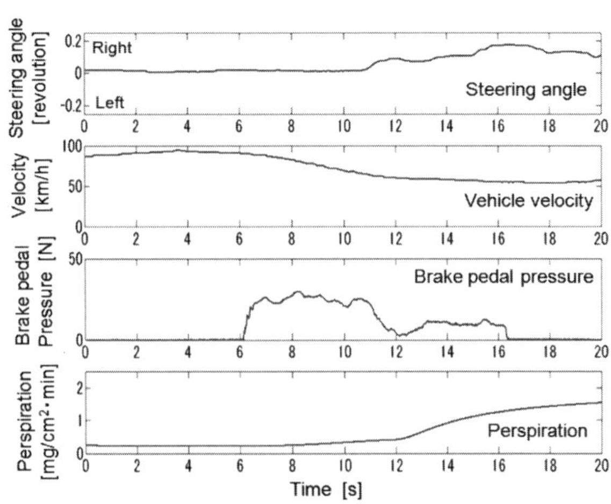

Fig. 4.8 Sample of driving signals recorded on exressway (steering angle, vehicle velocity, brake pedal pressure, and amount of perspiration)

recorded under the above four task conditions on urban roads and two conditions on an expressway. Driving data without any tasks are recorded as references before, between, and after the tasks. Driver's resting heart rate is also recorded before and after data recording in a quiet room.

4.3.2 Examples of Driving Data

Figure 4.8 shows examples of driving signals of steering angle, vehicle velocity, brake pedal pressure, and amount of perspiration recorded on the expressway. The amount of perspiration increased when the driver changed lanes.

4.4 Conclusion and Future Work

This chapter summarized our on-going data collection of real-world driving signals in Nagoya University, Japan. The project involves international collaboration with the United States and Turkey [1, 2]. Data collection is currently underway [10], and to date 150 drivers have participated. We will continue data collection and focus on comparative analysis of the driving data collected in these three locations.

Acknowledgments This work is partially supported by a NEDO grant "International Research Coordination of Driving Behavior Signal Processing Based on Large Scale Real World Database."

References

1. H. Abut, H. Erdogan, et al., "Real-World Data Collection with 'UYANIK," Chapter 3 in this book.
2. P. Angkititrakul, J.H.L. Hansen, et al., "UTDrive: The Smart Vehicle Project," Chapter 5 in this book.
3. W. Li, T. Shinde, et al., "Multiple regression of log spectra for in-car speech recognition using multiple distributed microphones," IEICE Trans. Information and Systems, vol. E88-D, no.3, pp. 384–390, Mar. 2005.
4. H. Fujimura, C. Miyajima, K. Ito, and K. Takeda, "Analysis of a large in-car speech corpus and its application to the multimodel ASR," Proc. 2005 IEEE International Conference on Acoustics, Speech, and Signal Processing, vol. 1, pp. 445–448, Mar. 2005.
5. J.A. Healey and R.W. Picard, "Detecting stress during real world driving tasks using physiological sensors," IEEE Transactions of Intelligent Transportation Systems, vol. 6, no. 2, pp. 156–166, June 2005.
6. L. Malta, C. Miyajima, N. Kitaoka, and K. Takeda, "Multimodal estimation of a driver's affective state," Proc. 2008 Workshop on Affective Interaction in Natural Environments, Oct. 2008.
7. L. Malta, P. Angkititrakul, C. Miyajima, and K. Takeda, "Multi-modal real-world driving data collection, transcription, and integration using Bayesian network," Proc. 2007 IEEE Intelligent Vehicles Symposium, pp.150–155, June 2008.

8. H. Amata, C. Miyajima, A. Ozaki, T. Nishino, N. Kitaoka, and K. Takeda, "Abrupt steering detection based on the road construction ordinance and vehicle acceleration captured with drive recorders," Proc. 2008 IEEE International Conference of Innovative Computing Information and Control, June 2008.
9. L. Malta, C. Miyajima, and K. Takeda, "Analysis of drivers' responses under hazardous situations in vehicle traffic," Proc. 2007 IEEE Intelligent Vehicles Symposium, pp. 1144–1149, June 2007.
10. A. Ozaki, S. Hara, et al., "In-car speech data collection along with various multimodal signals," Proc. 2008 Language Resources and Evaluation Conference, May 2008.

5
UTDrive: The Smart Vehicle Project

Pongtep Angkititrakul, John H.L. Hansen, Sangjo Choi, Tyler Creek, Jeremy Hayes, Jeonghee Kim, Donggu Kwak, Levi T. Noecker, and Anhphuc Phan

Abstract This chapter presents research activities of UTDrive: the smart vehicle project, at the Center for Robust Speech Systems, University of Texas at Dallas. The objectives of the UTDrive project are to collect and research rich multi-modal data recorded in actual car environments for analyzing and modeling driver behavior. The models of driver behavior under normal and distracted driving conditions can be used to create improved in-vehicle human–machine interactive systems and reduce vehicle accidents on the road. The UTDrive corpus consists of audio, video, brake/gas pedal pressure, head distance, GPS information (e.g., position, velocity), and CAN-bus information (e.g., steering-wheel angle, brake position, throttle position, and vehicle speed). Here, we describe our in-vehicle data collection framework, data collection protocol, dialog and secondary task demands, data analysis, and preliminary experimental results. Finally, we discuss our proposed multi-layer data transcription procedure for in vehicle data collection and future research directions.

Keywords Driving-behavior signal · Driving distraction · Multi-modal driving corpus · UTDrive · CAN-bus · Microphone array · Automatic speech recognition · Route navigation dialog

5.1 Introduction

There has been significant interest in development of effective human–machine interactive systems in diverse environmental conditions. Voice-based route navigation, entertainment control, and information access (voice mail, etc.) represent domains for voice dialog systems in the car environment. In-vehicle speech-based interactive systems should allow the driver to stay focused on the road. Several studies [3, 6] have shown that drivers can achieve better and safer

P. Angkititrakul (✉)
Center for Robust Speech Systems (CRSS), Erik Jonsson School of Engineering and Computer Science, University of Texas at Dallas, Richardson, Texas, USA
e-mail: john.hansen@utdallas.edu

K. Takeda et al. (eds.), *In-Vehicle Corpus and Signal Processing for Driver Behavior*, DOI 10.1007/978-0-387-79582-9_5,

driving performance while using speech interactive systems to operate an in-vehicle system compared to manual interfaces. Although better interfaces can be incorporated, operating a speech interactive system will still divert a driver's attention from the primary driving task with varying degrees of distraction. Ideally, drivers should pay primary attention to operating the vehicle versus managing other secondary tasks that are not immediately relevant to the primary driving task.

With current life styles and advancing in-vehicle technology, it is inevitable that drivers will perform secondary tasks or operate driver assistance and entertainment systems while driving. In general, the common tasks such as operating the speech interactive systems in a driving environment include cell-phone dialing, navigation/destination interaction, e-mail processing, music retrieval, and generic command and control for in-vehicle telematics system. If such secondary tasks or distractions fall within the limit of the amount of spare cognitive load for the driver, he/she can still focus on driving.

Therefore, the design of safe speech interactive systems for in-vehicle environments should take into account factors from the driver's cognitive capacity, driving skills, and their degree of proficiency for the cognitive application load. With knowledge of human factors, an effective driver behavior model with real-time driving signals can be integrated into a smart vehicle to support or control driver assistance systems to manage driver distractions (e.g., suspend or adapt applications in a situation of heavy driving workload, provide alert if the distraction level is higher than a safety threshold).

Driving is a multi-task activity that comprises both the discrete and continuous nature of a driver's actions to manage various driving and non-driving tasks. Over the past several decades, modeling driver behavior has drawn much research attention. A number of studies have shown that driver behavior can be modeled and anticipated by the patterns of driver's control of steering angle, steering velocity, car velocity, and car acceleration [13], as well as the driver identity itself [5, 16]. Miyajima et al. [11] effectively employed spectral-based features of the raw pedal pressure signal with Gaussian mixture model (GMM) framework to model driver characteristics. Building an effective driver behavior recognition framework requires a thorough understanding of human behavior and the construction of a mathematical model capable of both explaining and predicting the drivers' behavioral characteristics.

In recent studies, several researchers have defined different performance measures to understand driving characteristics and to evaluate their studies. Such measures include driving performance, driver behavior, task performance, or others. Driving performance measures consist of driver inputs to the vehicle or measurements of how well the vehicle was driven along its intended path [2]. Driving performance measures can be defined by longitudinal velocity and acceleration, standard deviation of steering-wheel angle and its velocity, standard deviation of the vehicle's lateral position (lane keeping), mean following distance (or head distance), response time to brake, to name only a few. Driver behavior measures can be defined by glance time, number of glances, or awareness of drivers. Task performance measures can be defined by the time to complete a task and the quality of the completed task (e.g., do drivers acquire information they need from cell-phone calling). Therefore, synchronized multi-modal data acquisition is very important in these studies.

UTDrive is part of a NEDO-supported international collaboration between universities in Japan (Nagoya Univ.) [12], Italy (Univ. of Torino), Singapore (Nanyang Univ.), Turkey (Sabancı and Koç Univ.) [11], and USA (UT-Dallas). The UTDrive (USA) project has been designed specifically to

- Collect rich multi-modal data recorded in a car environment (i.e., audio, video, gas/brake pedal pressures, following distance, GPS information, and CAN-bus information including vehicle speed, steering degree, engine speed, and brake position),
- Assess the effect of human–machine interactive system on driver behavior,
- Formulate better algorithms to improve robustness of in-vehicle speech/ASR systems,
- Design adaptive dialog management which is capable of adjusting itself to support a driver's cognitive capacity, and
- Develop a framework for smart inter-vehicle communications.

The outcomes of this project will help to develop a framework for building effective models of driver behavior and driver to machine interactions for safe driving. This chapter is organized as follows. Section 5.2 discusses details of our multi-modal data acquisition in actual car-driving environment. Section 5.3 describes the data collection protocol of the UTDrive project. Section 5.4 is devoted to the driving signals. Driver distraction is discussed in Section 5.5. Section 5.6 concentrates on driver behavior modeling. Section 5.7 discusses the proposed multi-layer transcription framework for the in-vehicle corpus. Finally, Section 5.8 concludes the chapter with future work.

5.2 Multi-Modal Data Acquisition

In this section, we describe an overview of the UTDrive hardware setup for multi-modal data acquisition.

5.2.1 Audio

A custom-designed five-channel microphone array with omni-directional Knowles microphones was constructed on top of the windshield next to the sunlight visors to capture audio signals inside the vehicle. Each microphone was mounted in a small movable box individually attached to an optical rail, as shown in Fig. 5.1. This particular design allows the spacing between each microphone component to be adjustable along various scales (e.g., linear,

Fig. 5.1 Custom-designed adjustable-spaced five-channel microphone array

logarithmic) across the width of the windshield. In addition, the driver speech signal is also captured by a close-talk microphone, which provides a reference audio from the speaker (driver) and allows the driver to move his/her head freely while driving the data collection vehicle.

Since there are a variety of noises (e.g., air conditioning (A/C), engine, turn signals, and passing vehicles) present in the driving environment, the microphone array configuration allows us to apply beam-forming algorithms to enhance the quality of the input speech signals [9, 17]. More importantly, drivers have to modify their vocal effort to overcome such perceived noise levels, namely the Lombard effect [7, 10]. Such effects on speech production (e.g., speech under stress) can degrade the performance of automatic speech recognition (ASR) system more than the ambient noise itself [8]. At a higher level, interacting with an ASR system when focused on driving (e.g., cognitive load) may result in a speaker missing audio prompts, using incomplete grammar, adding extra pauses or fillers, or extended time delays in a dialog system. Desirable dialog management should be able to employ multi-modal information to handle errors and adapt its context depending on the driving situations and driver's cognitive capacity.

5.2.2 Video

Two Firewire cameras are used to capture visual information of the driver's face region and front view of the vehicle, as shown in Fig. 5.2. Visual cues of driver characteristics such as head movement, mouth shape, and eye glance are essential for studying driver behavior. In addition, several studies have shown that combining audio and video information from the driver can improve ASR accuracy for low SNR speech [4, 18]. Integrating both visual and audio contents allows us to reject unintended speech prior to speech recognition and significantly improves in-vehicle human–machine dialog system performance [18] (e.g., determining the movement of the driver's mouth, body, and head positions).

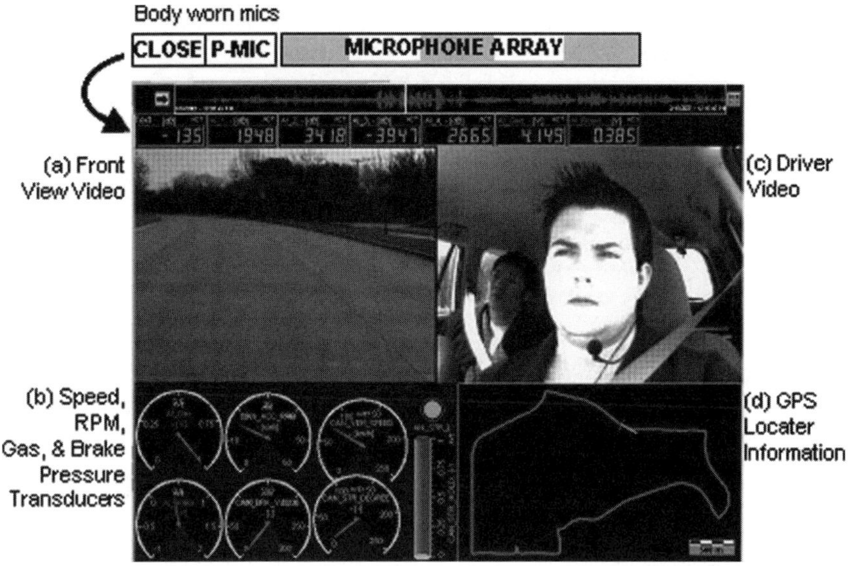

Fig. 5.2 A snapshot from a recording screen

5.2.3 CAN-Bus Information

As automotive electronics advance and government-required standards evolve, control devices that meet these requirements have been embracing modern vehicle design, resulting in the deployment of a number of electronic control systems. The controller area network (CAN) is a serial, asynchronous, multi-master communications protocol suited for networking vehicle's electronic control systems, sensors, and actuators. The CAN-bus signal contains real-time vehicle information in the form of messages integrating many modules, which interact with the environment and process high- and low-speed information. In the UTDrive project, we obtain the CAN signals from the OBD-2 port through the 16-point J1962 connection. Information captured from the CAN signals while the driver is operating the vehicle (e.g., steering-wheel angle, brake position, engine speed (RPM), and vehicle speed) is desirable in studying driver behavior.

5.2.4 Transducers and Extensive Components

In addition, the following transducers and sensors are included in the UTDrive framework (illustrated in Fig. 5.3):

- *Brake and gas pedal pressure sensors*: provide continuous measurement of pressure the driver puts on the pedals.

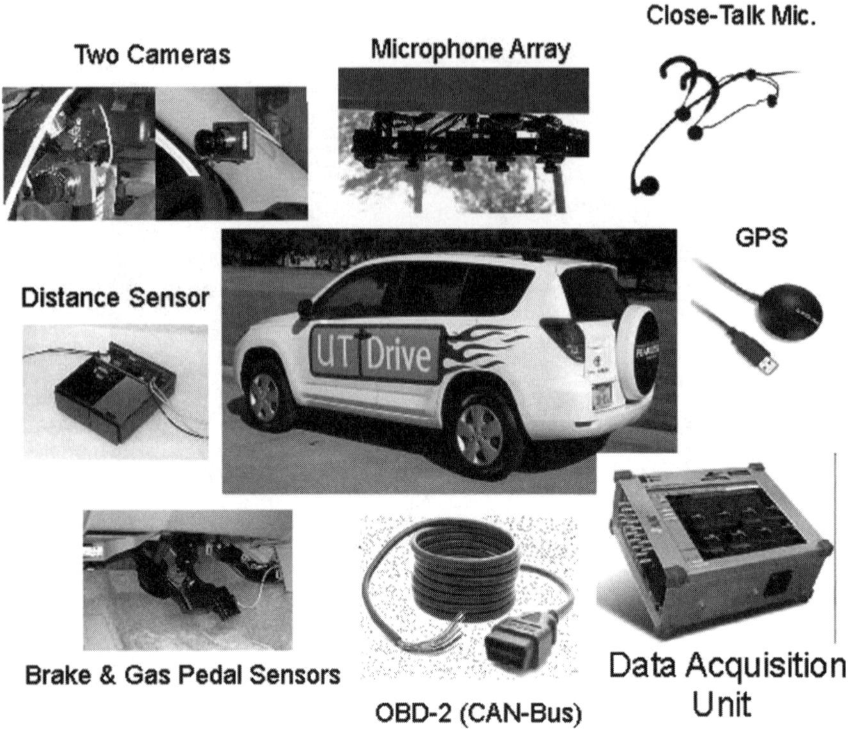

Fig. 5.3 UTDrive vehicle and its components

- *Distance sensor*: provides the head (or following) distance to the next vehicle.
- *GPS*: provides standard time and position/location of the moving vehicle.
- *Hands-free car kit*: provides safety during data collection and allows audio signals from both transmission sides to be recorded.
- *Biometrics*: heart-rate and blood pressure measurement.

5.2.5 Data Acquisition Unit (DAC)

The key component of effective multi-modal data collection is synchronization of the data. In our data collection, we use a fully integrated commercial data acquisition unit. With a very high sampling rate of 100 MHz, the DAC is capable of synchronously recording multi-range input data (i.e., 16 analog inputs, 2 CAN-bus interfaces, 8 digital inputs, 2 encoders, and 2 video cameras) and yet allows an acquisition rate for each input channel to be set individually. The DAC can also export all recording data as a video clip in one output screen or individual data in its proper format with synchronous time stamps.

The output video stream can be encoded to reduce its size and then transcribed and segmented with an annotation tool. Figure 5.2 shows a snapshot of a recording video clip with all data displayed on the screen (e.g., audio channels on top, two camera screens in the middle, sensors and CAN-bus information on the left bottom, and GPS information on the right bottom).

In order to avoid signal interference, both power cables and signal cables were wired separately along opposite sides of the car. The data acquisition unit is mounted on a customized platform installed on the backseat behind the driver. The power inverter and supply units are designed to be housed in the trunk space. Figure 5.3 shows the UTDrive data collection vehicle and its components.

5.3 Data Collection Protocol

For the data collection protocol, each participant drives the UTDrive vehicle using two different routes in the neighborhood areas of Richardson-Dallas, TX. The first route represents a residential area environment and the second route represents a business-district environment. Each route takes 10–15 min. The participant drives the vehicle along each route twice: the first being neutral/normal driving and the second being driving while performing secondary tasks.

Due to safety concerns, the assigned tasks are common tasks with mild to moderate degrees of cognitive load (e.g., interacting with commercial automatic speech recognition (ASR) dialog system, reading signs on the street, tuning the radio, having a conversation with the passenger, reporting activities, changing lanes).

The participants are encouraged to drive the vehicle up to three sessions of data collection with at least 1 week separation between sessions, in order to achieve session-to-session variability. Figure 5.4 shows the maps of two driving routes and the assigned tasks.

The assigned tasks are performed along each individual street/leg of the route and are alternated for three driving sessions. For example, a driver is requested to interact with a commercial voice portal while driving on one leg of the entire first session route. For the second and third sessions, the driver is asked to interact with another commercial voice portal and engage in conversation with the passenger while driving along the same leg of the entire route, respectively. This will allow us to compare different distraction levels with constant route driving conditions.

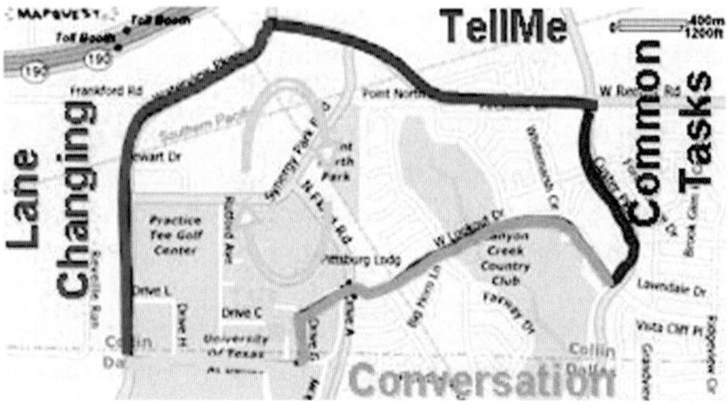

(a) Route 1: Business district.

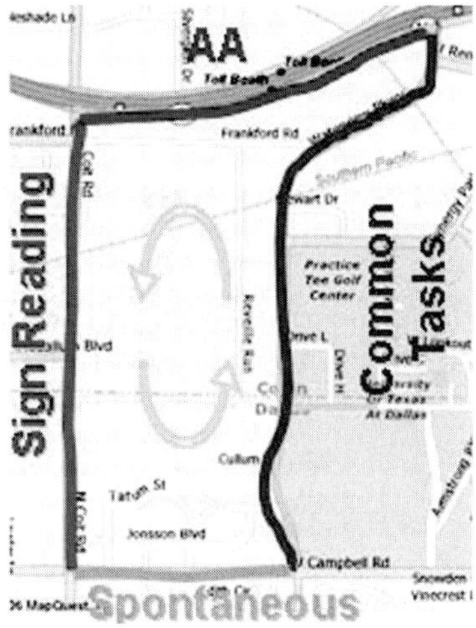

(b) Route 2: Residential district.

Fig. 5.4 The driving route 1 and the assigned tasks

5.4 Driving Signals

A variety of observable driving signals and sensory data have been applied to analyze and characterize driver behavior; for example, brake and gas pedal pressures, steering-wheel degree, velocity of vehicle, velocity of vehicle in front, acceleration, engine speed, lateral position, following distance, yaw angle (the angle between a vehicle's heading and a reference heading) are several presently under evaluation.

Our preliminary study focuses on four driving signals extracted from the CAN-bus information: acceleration (RPM), brake position, steering-wheel degree, and vehicle speed. Figure 5.5 shows plots, in normalized scales, of these four driving signals for 5 min of driving. Positive slope of steering degree

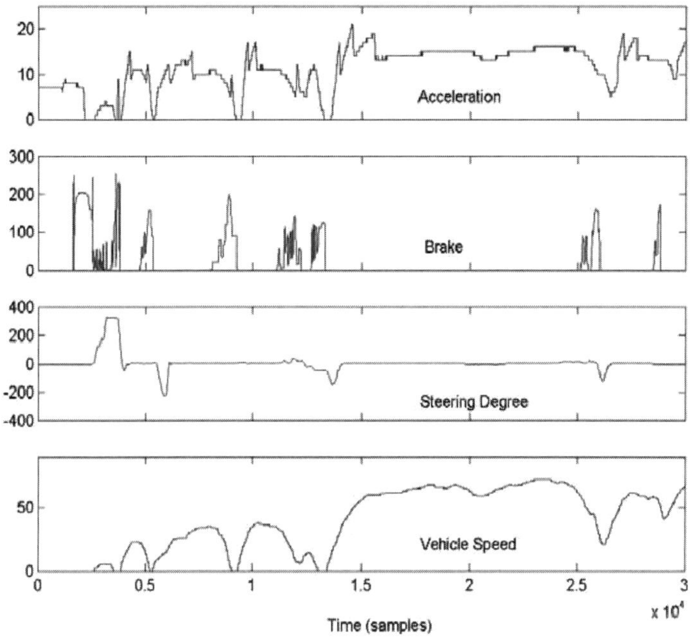

Fig. 5.5 Vehicle CAN-bus signals

plot represents counter-clockwise steering movement, while negative slope represents a clockwise steering maneuver.

5.5 Driver Distractions

Driver awareness has been a major safety concern since the invention of the automobile. According to the National Highway Traffic Safety Administration (NHTSA), there are four distinct types of driver distraction: visual, auditory, bio-mechanical (physical), and cognitive. Although these four modes of distraction are separately classified, they are not mutually exclusive.

For example, operating a mobile phone while driving may include all four types of driver distraction: dialing the phone (physical distraction), looking at the phone (visual distraction), holding a conversation (auditory distraction), and focusing on the conversation topic (cognitive distraction) [14].

Common sources of driver distraction are eating or drinking, focusing on other objects off the road, adjusting the radio, talking with passengers, moving objects in the vehicle, dialing and talking on a cell-phone, and others.

One approach to distraction detection is based on measurements of driving performance including steering-wheel movement, lateral lane position, longitudinal speed, lateral and longitudinal of acceleration and velocity, following

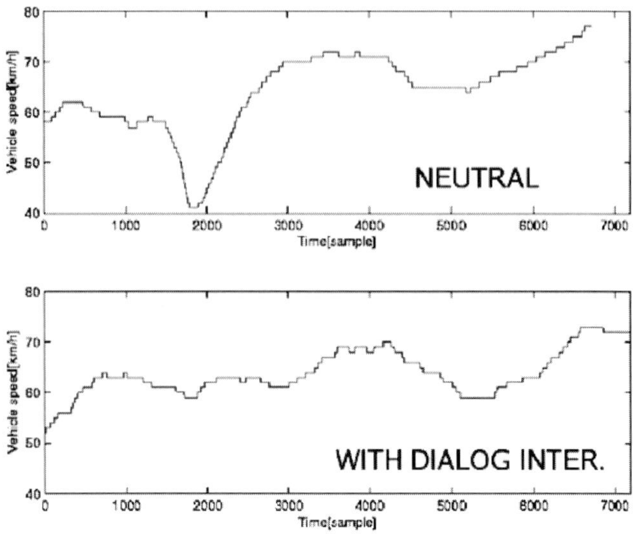

(a) Vehicle speed: 69.91 and 63.92 km/h average vehicle speeds with neutral driving and under distraction, respectively.

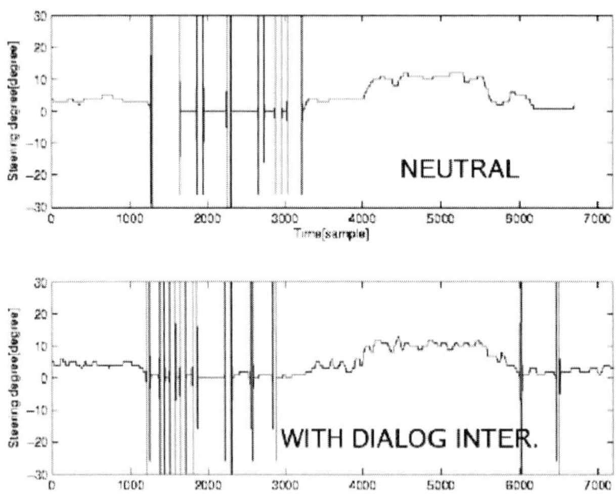

(b) Steering degree: 0.27 and 0.82 represent normalized short-term variances with neutral driving and under distraction, respectively.

Fig. 5.6 Comparison of neutral and distracted driving when the same drivers operate the vehicle on the same road (under very light traffic) with distraction-free (neutral) and with cell-phone-based voice dialog task interaction

distance (e.g., head distance), vehicle braking, and response time. Under distracted driving, drivers are likely to migrate away from smooth driving patterns (e.g., slow down or speed up vehicle speed, make excessive steering-wheel corrections for lane keeping).

Figure 5.6 shows plots of (a) vehicle speed and (b) normalized steering-wheel angle of a driver on the same route twice (under very light traffic). The neutral driving (do nothing—no distraction) is shown at the top of each plot, and the driving while interacting with a spoken dialog system is shown at the bottom. The vertical lines in plot (b) illustrate sharp corrections of the steering wheel between left and right. As we can see, the driver maintains a smoother driving pattern under the neutral condition versus when a secondary distraction task is present.

5.6 Driver Behavior Modeling

Driver behavior consists of lower-level components (e.g., eye movement and steering degree during lane keeping and lane changing) and higher-level cognitive components (e.g., maintaining situation awareness, determining strategies for navigation, managing other tasks) [15]. Therefore, effective modeling of driver behavior requires multidisciplinary knowledge of signal processing, control theory, information theory, cognitive psychology, physiology, and machine learning.

A driver behavior model can be developed to characterize different aspects of the driving tasks. For example,

- *Action classification/prediction*: Driver behavior model can be used to predict and categorize driver long-term behaviors such as turning, lane changing, stopping, and normal driving.
- *Driver verification/identification*: The goal here would be to recognize the driver by his/her driving-behavior characteristics.
- *Distraction detection*: The objective here is to identify whether the driver is under distraction due to performance of secondary tasks.
- *Route detection*: The goal here is to remove fine variation in the driving signals and extract the overall point-to-point route from origin to final destination.

5.7 Transcription Convention

One of the major challenges facing our efforts to utilize rich multi-modal data is a unified transcription protocol. Such protocols do not exist in the vehicle/driver modeling community. Multi-layer transcription is necessary for this study. For example,

- *Audio*: different types of background noise inside and outside the vehicle, passengers' speech, radio and music, ring tone, and other audio noise types.
- *Driving environment*: type of roads (number of lanes, curve or straight, highway or local, speed limit), traffic (traffic situation, traffic light, surrounding vehicles), road condition, etc.
- *Driver activity*: look away from the road, talk to passengers, dial-a-phone, talk on a phone, look at rear mirror, look at control panel, sleepy, daydreaming, etc.
- *Vehicle mode*: left or right turn, left or right lane change, U-turn, stop and go, stop, etc.

The ability to formulate an effective transcription convention is critical in developing future directions for smart vehicle research. The transcription convention used will lead to better algorithm development which reduces the cognitive loads on the driver for smart vehicle systems.

5.8 Conclusion and Future Work

This chapter has described research activities of the UTDrive project [19] and our vehicle setup for real-time multi-modal data acquisition in an actual driving environment. Further details on advances in UTDrive can be found at [19]. Example profiles using analysis of CAN-bus information illustrates the range of research possible with the UTDrive corpus. However, robust and reliable driver-behavior modeling systems need to employ other modalities of data such as video and driver's biometric information to better integrate the driver and system designs of the future.

While a range of research opportunities are possible, focused efforts on modeling driver behavior when additional technology is present (e.g., route navigation, cell-phone interactive, dialog systems) will allow for advances in next-generation vehicle systems which are safer with reduced cognitive task loads.

Acknowledgments This work is supported by NEDO, Japan and University of Texas at Dallas under project EMMITT. The authors would like to thank all the student assistants for their great contributions during this project.

References

1. H. Abut, H. Erdogan, A. Ercil et al., "Data collection with 'UYANIK': too much pain; but gains are coming," Chapter 3 in this book.
2. A. Baron, P. Green, "Safety and usability of speech interfaces for in-vehicle tasks while driving: a brief literature review," *UMTRI-2006-5: The University of Michigan, Transportation Research Institute*, Ann Arbor, February 2006, pp.1–8.

3. C. Carter, R. Graham, "Experimental comparison of manual and voice controls for the operation of in-vehicle systems," in *Proceedings of the IEA2000/HFES2000 Congress*, Santa Monica, CA.
4. T. Chen, "Audio-visual speech processing," *IEEE Signal Process Mag.*, vol. 18, no. 1, pp. 9–21, 2001.
5. H. Erdogan, A. Ercil, H.K. Ekenel, S.Y. Bilgin, I. Eden, M. Kirisci, H. Abut, "Multi-modal person recognition for vehicular applications," N.C. Oza et al. (Eds.) *MCS-2005*, Monterey, CA, LNCS-3541, pp. 366–375, Jun. 2005.
6. C. Forlines, B. Schmidt-Nielsen, B. Raj, P. Wittenburg, P. Wolf, "Comparison between spoken queries and menu-based interfaces for in-car digital music selection," *TR2005-020*, Mitsubishi Electric Research Laboratories, Cambridge, MA.
7. J.H.L. Hansen, D. Cairns, "ICARUS: a source generator based real-time system for speech recognition in noise, stress, and Lombard effect," *Speech Commun.*, vol. 16, no. 4, pp. 391–422, July 1995.
8. J.H.L. Hansen, "Analysis and compensation of speech under stress and noise for environmental robustness in speech recognition," *Speech Commun.: Special Issue on Speech Under Stress*, vol. 20, no. 2, pp. 151–170, Nov. 1996.
9. T.B. Hughes, H.S. Kim, J.H. DiBiase, H.F. Silverman, "Performance of an HMM speech recognizer using a real-time tracking microphone array as input," *IEEE Trans. Speech Audio Process.*, vol. 7, no. 3, pp. 346–349, 1999.
10. E. Lombard, "Le signe de l'elevation de la voix," *Ann. Maladies Oreille Larynx*, Nez, Pharynx, vol. 37, pp. 101–119, 1911.
11. C. Miyajima, Y. Nishiwaki, K. Ozawa, T. Wakita, K. Itou, K. Takeda, F. Itakura, "Analysis and modeling of personality in driving behavior and its application to driver identification," *Proc. IEEE*, vol. 95, no. 2, pp. 427–437, Feb. 2007.
12. C. Miyajima, T. Kusakawa, T. Nishino, N. Kitaoka, K. Itou, K. Takeda, "On-going data collection for driving behavior signal," Chapter 4 in this book.
13. A. Pentland, A. Liu, "Modeling and prediction of human behavior," *Neural Comput.*, vol. 11, pp. 229–242, 1999.
14. M. Pettitt, G. Burnett, A. Stevens, "Defining driver distraction," *World Congress on Intelligent Transport Systems*, San Francisco, Nov. 2005.
15. D. Salvucci, E.R. Boer, A. Liu, "Toward an integrated model of driver behavior in a cognitive architecture," *Transport. Res. Rec.*, no. 1779, pp. 9–16, 2001.
16. A. Wahab, T.C. Keong, H. Abut, K. Takeda, "Driver recognition system using FNN and statistical methods," H. Abut, J.H.L. Hansen, K. Takeda (Eds.) Chapter 3 in *Advances for in-vehicle and mobile systems*, Springer, New York, 2007.
17. X.X. Zhang and J.H.L. Hansen, "CSA-BF: a constrained switched adaptive beamformer for speech enhancement and recognition in real car environment," *IEEE Trans. Speech Audio Proc.*, vol. 11, no. 6, pp. 733–745, Nov. 2004.
18. X.X. Zhang, K. Takeda, J.H.L. Hansen, T. Maeno, "Audio-visual speaker localization for car navigation systems," in *INTERSPEECH-2004*, Jeju Island, Korea, 2004.
19. www.utdallas.edu/research/utdrive.

6
Wireless Lan-Based Vehicular Location Information Processing

Seigo Ito and Nobuo Kawaguchi

Abstract This chapter focuses on environmental signals like wireless signals which can observe from outside of vehicle. These signals are mostly used for localization of a terminal in mobile system. Particularly, we examine the wireless LAN-based localization in vehicles. Additionally, we explore the possibility of orientation estimation in vehicles using wireless LAN. Vehicles themselves interrupt wireless signals and act as a big obstacle. This causes a difference in signal strength distribution according to the location of wireless LAN antenna. By using these differences we can estimate vehicles' orientation. Finally, we introduce our metropolitan-scale localization project named Locky.jp.

Keywords Wireless LAN · Location information processing · Orientation estimation · Wireless location systems · Metropolitan-scale location services · Localization technologies · Signal strength distribution

6.1 Introduction

For the research and development of driver support system, many types of data have been collected for a vehicle, for example, driving signal (e.g., gas pedal, brake pedal), car state signal (e.g., speed), car location (e.g., inter-vehicular distance), video signal, and sound signal. In addition to these signals, we focused on environmental signals, like wireless signals, which is emitted outside the vehicles. When driving vehicles we can observe various wireless signals, for example, GSM (global system for mobile communications) cell towers signals, FM–AM radio signals, GPS (global positioning system), and wireless LAN signals. It is possible to consider these infrastructures as a type of lighthouse. These environmental signals are very useful for various purposes. The most major use is localization. As another usage of these signals, for example, Krumm proposed destination estimation method [1] named predestination,

S. Ito (✉)
Graduate School of Engineering, Nagoya University, Nagoya, Japan

K. Takeda et al. (eds.), *In-Vehicle Corpus and Signal Processing for Driver Behavior*, DOI 10.1007/978-0-387-79582-9_6,

using vehicles trajectories along with data about driving behaviors. Sohn proposed mobility detection method [2] using everyday GSM signals. This method estimates user's mobility by comparing difference in signal strength distribution among some mobilities. Our results show that environmental wireless signals can be used for various purposes [9].

We focused especially on wireless LAN signals. Due to the widespread use of wireless LAN in these days, we can observe wireless LAN signals everywhere. And almost all laptops and PDAs have wireless LAN adapter. It is easy to collect data with the existing vehicular system.

This chapter is composed of three parts. First, we introduce and examine wireless LAN-based signal processing for localization. Second, we explore the possibility of wireless LAN-based signal processing for orientation estimation. Finally, we introduce our metropolitan-scale localization project named Locky.jp and present conclusions.

6.2 Localization

6.2.1 Wireless LAN-Based Localization

For explaining wireless LAN-based localization, we introduce the scheme of wireless LAN-based localization. In wireless LAN, many access points emit beacons periodically. Wireless terminals can obtain various information from these beacons, such as SSID, MAC address of APs (BSSID), and signal strength. If a vehicle enters a wireless LAN-available area, the vehicle can get these beacons', information. Then, if the access point's position is known, the vehicle can estimate its position in relation to the access point. This is the basic scheme of wireless LAN-based localization (Fig. 6.1).

Wireless LAN-based localization can be mainly classified into proximity approach [3], triangulation approach [4, 5] and scene analysis approach [6, 7, 8].

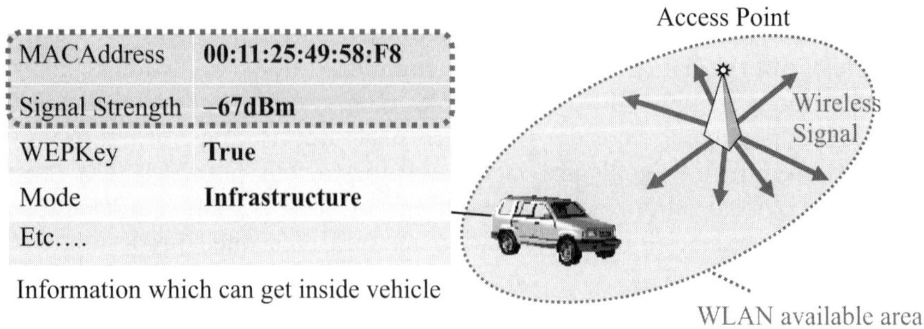

Fig. 6.1 Scheme of WLAN localization

6.2.2 Proximity Approach

The proximity approach is the easiest way for localization. This method considers the nearest access point position as a terminal's position. When the terminal has position list of access points, the terminal's can estimate its position by proximity easily. However, the accuracy of this method is low.

6.2.3 Triangulation Approach

The triangulation approach uses reference points whose position is known (e.g., access point). One type of this approach is also used by GPS. When systems use relative distance from a reference point, it is called lateration. Figure 6.2. shows overview of lateration. When a terminal observes signal strength (ss1, …, ssn) from known access points [($x1$,$y1$), …, (xn,yn)], the system calculates the terminal's position (xt,yt) as follows (n is the number of access points which the user observed):

$$(x_t, y_t) = \left(\frac{\sum_{i=1}^{n} \frac{x_i}{10^{-32-\mathrm{ss}_i/-32-\mathrm{ss}_i 25}}}{\sum_{i=1}^{n} \frac{1}{10^{-32-\mathrm{ss}_i/-32-\mathrm{ss}_i 25}}}, \frac{\sum_{i=1}^{n} \frac{y_i}{10^{-32-\mathrm{ss}_i/-32-\mathrm{ss}_i 25}}}{\sum_{i=1}^{n} \frac{1}{10^{-32-\mathrm{ss}_i/-32-\mathrm{ss}_i 25}}} \right)$$

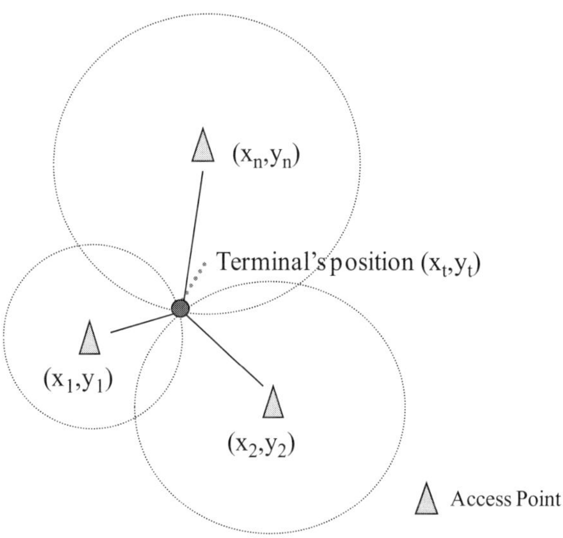

Fig. 6.2 The triangulation approach (lateration)

6.2.4 Scene Analysis Approach

This approach uses prior knowledge of the environment like a radio map. This approach has high accuracy. However, to get high accuracy, this approach needs fine-grained pre-sampling of the environment. For example, if a system needs 1 ft accuracy, approximate equivalent interval's pre-sampling is required. Therefore, it is not suitable for vehicular localization. Our previous work [6] using scene analysis shows 64% of localization requests are within 2 m in indoor environment.

6.2.5 Accuracy in Outdoor Configuration

To examine wireless LAN-based localization accuracy outdoors, we collected information of access points (Fig. 6.3). Collectors of access point information carried a laptop, wireless LAN adapter, wireless survey software, and GPS in a backpack. By using this backpack they acquired the access point's position.

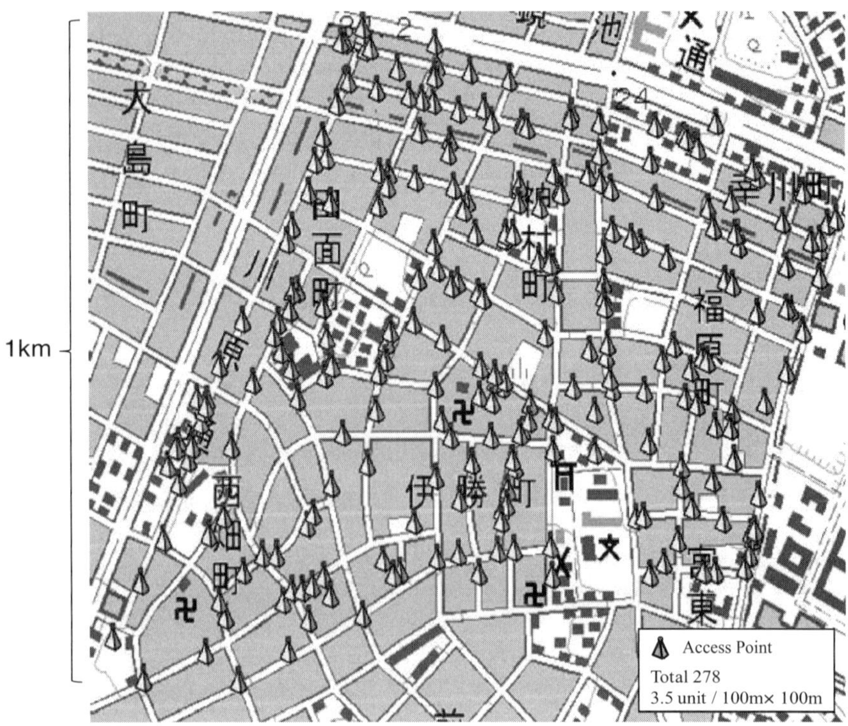

Fig. 6.3 Access point map (residential area)

Table 6.1 Accuracy of WLAN localization in outdoors

Vehicle	Average (m)	SD (m)	Coverage (%)
Bicycle	25.4	21.2	98.3
Motorbike	27.7	18.8	97.2
Car	27.4	23.8	96.7

After collecting access point's position, we examined the accuracy of wireless LAN-based localization.

We considered GPS position as correct position and calculated wireless LAN-based positioning accuracy. We conducted experiments with three types of vehicles: a bicycle, a motorbike, and a car. Table 6.1 shows the result, accuracy average, standard deviation, and coverage. Coverage means the ratio using which vehicles estimate their position using only wireless LAN. There was a little difference of coverage among the three vehicles due to their differences in speed. We empirically observed that according to the increase in vehicle speed, the number of access points which vehicles can observe is decreasing. Therefore, it is considered that the bicycle has high coverage and the car has low coverage in this area.

6.3 Orientation Estimation

6.3.1 Difference of Signal Strength

In the previous section, we introduced localization methods of wireless LAN. But it is empirically known that received signal strength varies according to the relative angles between terminals and access points.

For the purpose of survey, we observed the signal strength distributions for each orientation at same positions (Fig. 6.4). Figure 6.5 shows the average value of received signal strength for each orientation. In Fig. 6.5, the distance from the middle of a circle shows the average received signal strength, and each axis shows the relative angle between the terminal and the access point. We obviously see the difference of received signal strength for each orientation. For example, in wireless LAN adapter A, the terminal observed the highest average at relative angle 0 when the user is facing the access point. Even in the same location, received strength signal varies according to the wireless LAN adapter.

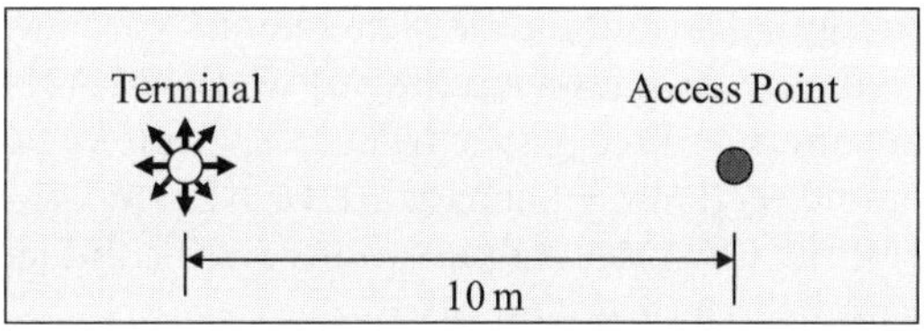

Fig. 6.4 Observation setting

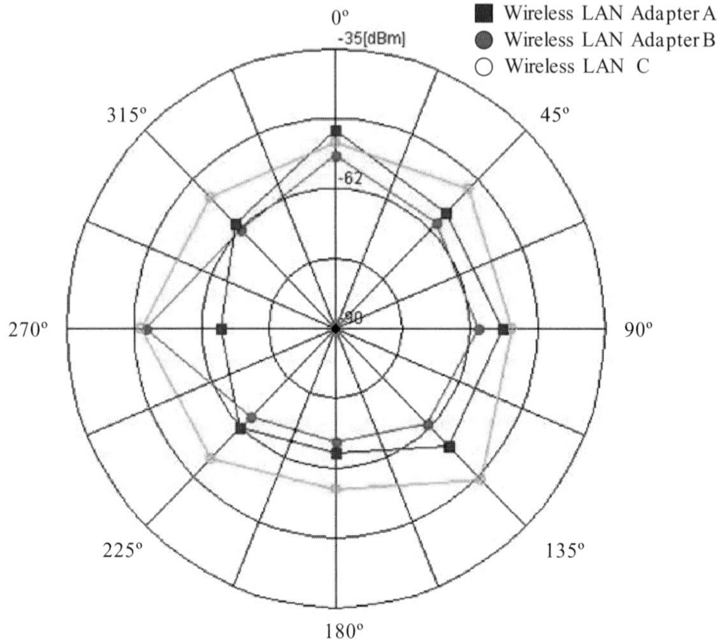

Fig. 6.5 Average signal strength for each orientation

By using these differences of received signal strength and Bayesian filter [9], we have proposed a wireless LAN-based orientation estimation method [10]. Our method compares signal strength distribution, uses KL-divergence [11] and estimates orientation. With our method, users can determine orientation only using a wireless LAN adapter. The accuracy of two-way estimation is 92% and four-way estimation is 83% under 6 s observation of four access points.

6.3.2 Orientation Estimation in Vehicle

When we collect wireless signal as an environmental signal in vehicle, we must consider the difference of signal strength distribution according to the antenna's location because the vehicle itself interrupts wireless signal and works like a big obstacle. If we think reverse, there is a possibility of orientation estimation according to the antennas location. We examine differences of received signal strength distribution for each antenna's location.

Figure 6.6 shows a setting of wireless LAN antennas. We set wireless LAN antennas in three places: front side, inside vehicle, and backside of vehicle. The omnidirectional antenna is BUFFALO WLE-NDR. The vehicle is TOYOTA Hybrid ESTIMA. Observations were conducted for 2 min to each location while the car was stopped.

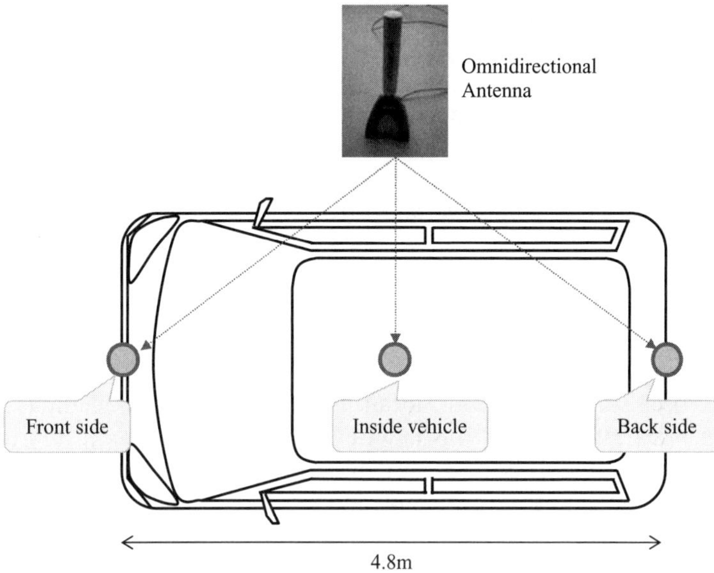

Fig 6.6 Antenna's location

Figure 6.7 shows the result of probability density of received signal strength from an access point at each antenna. *X*-axis shows the received signal strength and *Y*-axis shows probability density. At the backside and inside of the vehicle, averages were −48 and −50 dBm, respectively. In contrast, the average

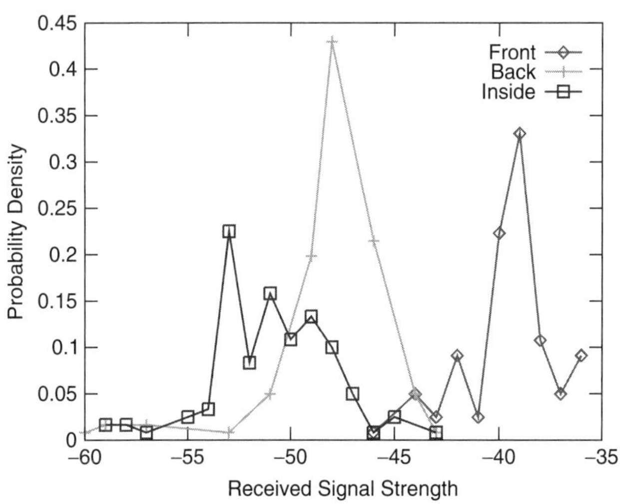

Fig. 6.7 Difference of signal strength

at front side was −39 dBm. These differences of received signal strength distribution are enough to apply our orientation estimation method [10] described in Section 6.3.1.

If we use wireless LAN for easy orientation sensor, there are various usage scenarios. For example, if a car stopped, a system cannot estimate the car's orientation-only GPS. However, if the system is equipped with a wireless LAN adapter, it is easy to estimate the car's orientation.

6.4 Metropolitan-Scale Localization

We have been studying the basic localization and orientation methods. However, for practical purposes, it is required to collect access point information widely and to expand the service-available area. Therefore, we have surveyed a major city in Japan and have attempted to evaluate the feasibility of metropolitan-scale localization based on wireless LAN. In addition, we now introduce metropolitan-scale localization project Locky.jp which uses the environmental signals as a reference point for localization.

6.4.1 Feasibility of Metropolitan-Scale Localization

For the purpose of feasibility examination for metropolitan-scale localization, we conducted survey and accuracy evaluation in major cities (Fig. 6.8). We surveyed 1 km^2 of commercial area in Tokyo, Osaka, and Nagoya. In this survey, we used wireless survey backpack described in Section 6.2.5.

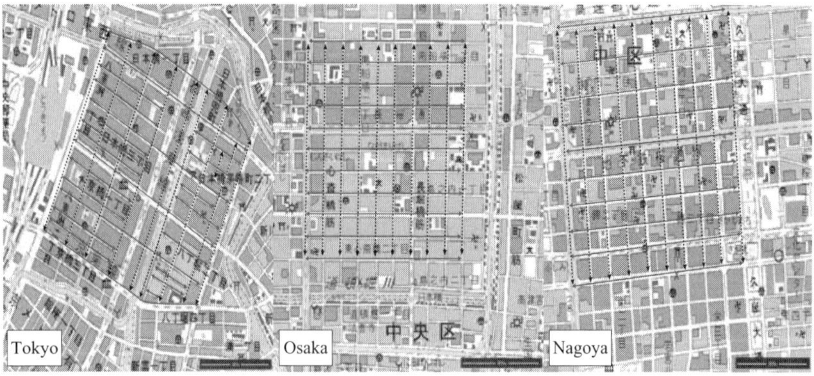

Fig. 6.8 Survey area

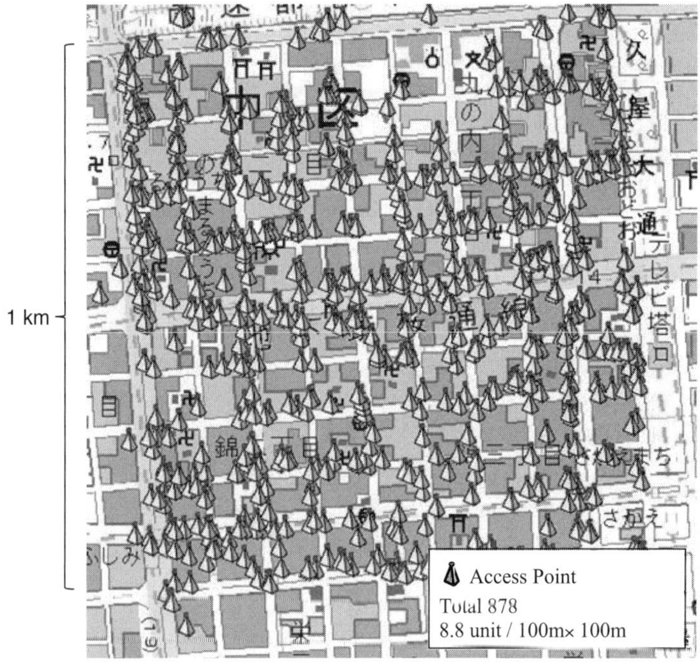

Fig. 6.9 AP map for Nagoya commercial district

As a result of the survey, we have found 2746 (Tokyo 928, Osaka 940, Nagoya 878) access points in these three areas, and there are full coverage of wireless LAN. Figure 6.9 shows the result of survey in Nagoya. By using this surveyed data, we have examined the estimation accuracy for each city. In this case, we have used the localization method described in Section 6.2.3.

Table 6.2 shows the result of accuracy, average, and coverage in these three cities. It should be noted that all cities have more than 97% coverage of wireless LAN-based localization. Even when a terminal cannot receive signal from access point, only a few meters move enables the terminal to observe wireless LAN signal. In other words, wireless LAN-based metropolitan-scale localization is available everywhere and has great possibility.

Table 6.2 Accuracy and coverage in city area

Vehicle	Average (m)	SD (m)	Coverage (%)
Tokyo	40.3	28.8	97.6
Osaka	34.7	25.8	98.1
Nagoya	48.7	33.5	98.8

6.4.2 Locky.jp

In the previous section, we have pointed out the feasibility of metropolitan-scale wireless LAN-based localization and conclude that metropolitan-scale wireless LAN-based localization has a great deal of potential. Therefore, we launched the project named Locky.jp to make wide-area wireless LAN-based localization come true.

Locky.jp [12] is a metropolitan-scale localization project designed to collect reference trace sets covering major cities in Japan among users collaboration. Reference trace set (RTS) consists of BSSID (basic service set identifier), signal strength (SS), and position of access point (latitude and longitude).

$$\mathrm{RTS} = \{\mathrm{BSSID},\ \mathrm{ss},\ \mathrm{latitude}, \mathrm{longitude}\}$$

Locky.jp has the following three main goals:

1. Collection of metropolitan-scale reference trace set in Japan (and also abroad). The realization of a metropolitan-scale positioning system needs at least a prior survey of the wireless LAN environment. Locky.jp aims to construct this system among users collaboration.
2. Provision of a client program for the positioning system based on 802.11. We provide the "Locky client" that can estimate the device location based on reference trace set from a large number of users.
3. Development and provision of location-aware application. The "Locky client" provides different types of information according to the location. Locky.jp is also aiming at the development of different location-aware application.

Figure 6.10 shows the overview of locky.jp. Volunteers collect wireless LAN reference trace set. By using them and the localization method we introduced in Section 6.2, the system estimates terminals' position. Then, reference traces set are uploaded to locky.jp server, and database is constructed. Finally, the database is released to users with some applications. To construct a wide-area positioning system, it is important to collect the reference trace set effectively and widely.

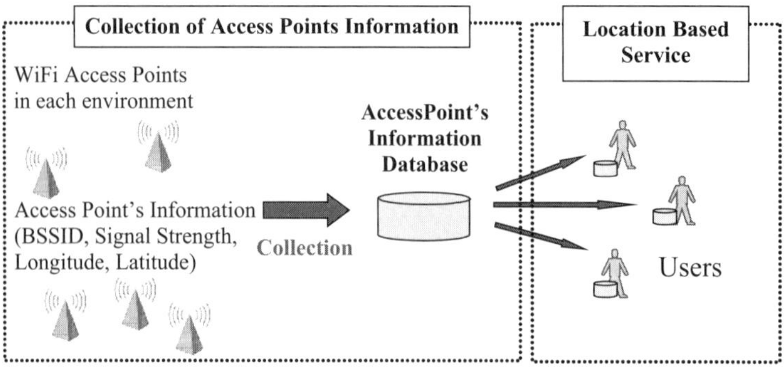

Fig. 6.10 Locky.jp overview

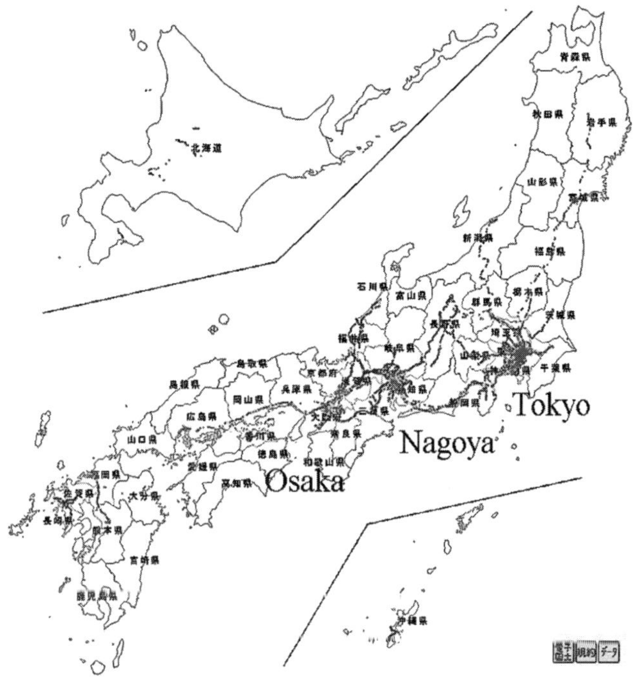

Fig. 6.11 Locky.jp access point map (Japan)

6.4.3 Current status

Figure 6.11 shows all reference trace sets as of May 11, 2007. We have collected 308,607 reference trace sets (the number of access points) with 185 collaborators. These collaborators consist of universities, companies, and many other ordinary persons. We provide wireless LAN survey software named Locky Stumbler. Collaborators survey reference trace sets using Locky Stumbler.

We also provide collected data as locky.jp database with the Java API. By using these database and API, developers can build functions of wireless LAN-based localization in their applications.

6.4.4 Related Projects

There are some projects that aim to construct metropolitan-scale wireless LAN-based localization. Among them "Place Engine" [13] provides a WEB API using wireless LAN-based localization. When a user installs a Place Engine client to his terminal, the client collects wireless LAN information and sends a positioning query to Place Engine server. On the other hand, "PlaceLab" [14] is a project to support development of a location information system by providing a toolkit for position estimation and a database for wireless beacon (IEEE802.11, GSM, and Bluetooth) information. PlaceLab's toolkit also supports collection of the reference trace set.

6.4.5 On-Going and Future Work

As a next step of Locky.jp, we are considering construction of reference trace set in the basement and hybrid usage with GPS both indoors and outdoors. In the basement, like underground city and subway, GPS is not available. However, wireless LAN signal is available in such environment. We are discussing about construction of reference trace set in such environments.

As an issue of wireless LAN-based metropolitan-scale positioning, there is refreshment of reference trace set. Position of access points may change, because of a crash, a move, or destruction, etc. Therefore, we examined an existence transition of access points. We had collected access points' information for 316 days in the same route. Figure 6.12 shows transitions of access points. *X*-axis shows elapsed days and *Y*-axis shows each access point. If an access point is observed through the route, the day is plotted. On the first day, we observed 382 unique access points. However, on the last day, we observed 223 out of 382 access points. In other words, 41.6% access points were not observed. Further study will be necessary to refresh reference trace set, for example, self-mapping method.

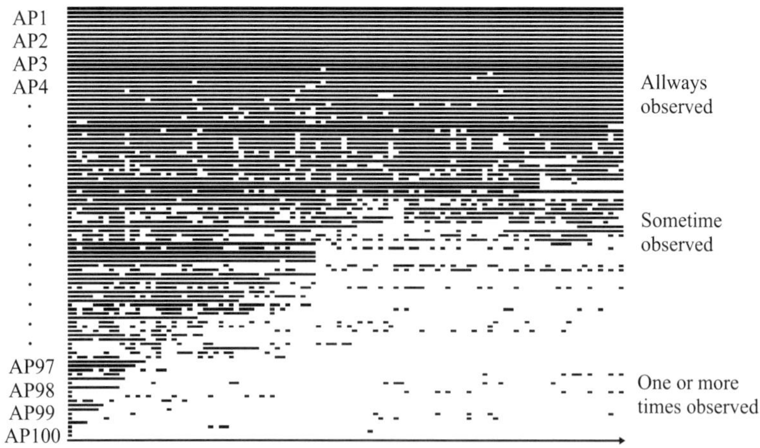

Fig. 6.12 Transition of access point existence

6.5 Conclusion

In this chapter, we focused on environmental signal and introduced wireless LAN-based localization and orientation estimation for vehicles. Additionally, we introduced metropolitan-scale localization project Locky.jp. As we described above, wireless LAN-based localization and orientation estimation has a great potential. We still continue to study the improvement of accuracy and coverage.

Furthermore, wireless LAN can be used for interesting applications. For future application, we are discussing about some applications with wireless LAN on vehicle. For example, we can use wireless LAN as an easy inter-vehicular distance sensor, a multi-hop car-to-car communication, a crash detection sensor, and a motion detection sensor. At close range, wireless LAN signal is sensitive; therefore, it is possible to use wireless LAN as sensor in these scenaris.

References

1. J. Krumm and E. Horvitz: Predestination: Inferring Destinations from Partial Trajectories. In Proceedings of the Eighth International Conference on Ubiquitous Computing (Ubicomp2006), 2006.
2. T. Sohn, A. Varshavsky, A. LaMarca, M. Y. Chen, T. Choudhury, S. Consolvo, J. Hightwer, W. G. Grisword, and E. de Lara: Mobility Detection Using Everyday GSM Trasces. In Proceedings of the Eighth International Conference on Ubiquitous Computing (Ubicomp2006), 2006.
3. K. Cheverst, N. Davies, K. Mitchell, and A. Friday: Experiences of Developing and Deploying a Context-Aware Tourist Guide: The GUIDE Project. In Proceedings of the Sixth Annual International Conferences on Mobile Computing and Networking (MOBICOM 2000), pp. 20–31, 2000.
4. T. Kitasuka, T. Nakanishi, and A. Fukuda: Wireless LAN based Indoor Positioning System WiPS and Its Simulation. In Proceeding of IEEE Pacific Rim Conference on Communications, Computers and Signal Processing (PACRIM'03), pp. 272–275, 2003.
5. Y. Iwasaki, N. Kawaguchi, and Y. Inagaki: Design, Implementation and Evaluations of a Direction Based Service System for Both Indoor and Outdoor. In Proceedings of the Second International Symposium on Ubiquitous Computing Systems (UCS 2004), pp. 7–14, 2004.
6. S. Ito and N. Kawaguchi: Bayesian based Location Estimation System Using Wireless LAN. In Third IEEE Conference on Pervasive Computing and Communications Workshops, pp. 273–278, (2005).
7. J. Hightower and G. Borriello: Particle Filters for Location Estimation in Ubiquitous Computing: A Case Study. In Proceedings of the Sixth International Conference on Ubiquitous Computing (Ubicomp2004), pp. 88–106, 2004.
8. V. Seshadri, G. V. Zaruba, M. Huber: A Bayesian Sampling Approach to In-door Localization of Wireless Devices Using Received Signal Strength Indication. In Proceedings of The Third IEEE International Conference on Pervasive Computing and Communications (PerCom2005), pp. 75–84, 2005.
9. S. Thrun, W. Burgard, and D. Fox : Probabilistic Robotics, The MIT Press, 2005.

10. S. Ito, N. Kawaguchi: Orientation Estimation Method Using Divergence of Signal Strength Distribution. In Third International Conference on Networked Sensing Systems, pp. 180–187, 2006.
11. I. Dagon, L. Lee, and F. C. N. Pereira: Similarity-Based Models of Word Co-occurrence Probabilities, Machine Learning, vol. 34, no. 1–3, pp. 43–69, 1999.
12. Locky Project : locky.jp/en/.
13. Place Engine : www.placeengine.com/.
14. A. LaMarca, Y. Chawathe, S. Consolvo, J. Hightower, I. Smith, J. Scott, T. Sohn, J. Howard, J. Hughes, F. Potter, J. Tabert, P. Powledge, G. Borriello, and B. Schilit : Place Lab: Device Positioning Using Radio Beacons in the Wild. In Proceedings of The Third International Conference on Pervasive Computing (PERVASIVE 2005), pp. 116–133, 2005.

7
Perceptually Optimized Packet Scheduling for Robust Real-Time Intervehicle Video Communications

Enrico Masala and Juan Carlos De Martin

Abstract The automotive industry is increasingly looking at solutions for intervehicle wireless communications, as demonstrated by the development of wireless access in vehicular environment (WAVE) communications standards. Potential applications are numerous, ranging from safety to entertainment. Warning signals are the most immediate applications, but more complex forms of communications, and video, in particular, could be used by innovative applications such as multi-vehicle-based visual processing of road information, multi-vehicle radar systems for obstacle avoidance and automatic driving, and more generally swarm communications among cars traveling along the same road. However, providing reliable communication services in the vehicular environment still remains a challenging task, mainly because of the extremely time-varying characteristics of wireless channels. This chapter presents a perceptually optimized packet scheduling algorithm developed for robust low-delay intervehicle video communication. The algorithm has been tested by transmitting video data captured by on board cameras to another vehicle in proximity, using ad hoc 802.11 wireless technology and the H.264 video coding standard, showing that it achieves a consistently higher quality compared to two reference techniques, i.e., the standard MAC-layer retransmission scheme and a delay-constrained retransmission technique, with gains up to 2 dB PSNR.

Keywords Ad hoc wireless networks · Intervehicle video communications · Perceptual optimization · Packet scheduling · Automatic repeat request

7.1 Introduction

The automotive industry is highly interested in solutions which can provide some form of driving assistance, even if it is not yet clear what is the best form to deliver it, as it is highlighted in Chapter 1. However, it is generally agreed that to

E. Masala (✉)
Dipartimento di Automatica e Informatica–Politecnico di Torino–Corso Duca degli Abruzzi 24, 10129, Torino, Italy
e-mail: masala@polito.it

K. Takeda et al. (eds.), *In-Vehicle Corpus and Signal Processing for Driver Behavior*, DOI 10.1007/978-0-387-79582-9_7,

achieve this objective vehicles need some sort of real-time context awareness, obtained for instance by communicating with the surrounding vehicles. Such communications might not only be aimed at improving safety, as it is currently expected, but in the end they might also contribute to a more pleasant driving experience.

In recent years increasing efforts have been devoted to address open issues in the field of intervehicle wireless communications. As mentioned, warning and safety signals are the most immediate applications, but more complex forms of communications, and video, in particular, could be used by innovative applications, such as multi-vehicle-based visual processing of road information aimed at, for instance, obstacle detection thus contributing to automatic driving. Other applications include generic communications among cars traveling along the same road, e.g., to enable conferencing applications among passengers.

Communications standards such as the wireless access in vehicular environment (WAVE) standard, i.e., the IEEE 1609.x family, are expected to facilitate intervehicle communications, addressing issues including security, management of multiple radio channels, and system resources. The IEEE 802.11p standard, an extension of the well-known 802.11, is under active development and it is expected to cover protocol and networking services in WAVE.

In the meanwhile, several researchers have performed vehicle-to-vehicle and vehicle-to-roadside communication experiments based on the 802.11 standard [1, 2], mainly focusing on throughput and connectivity issues. Others investigated routing [3] and information dissemination issues [4]. Modifications of the existing MAC protocols have also been proposed [5] to better match the needs of intervehicle communications such as, for instance, reliable channel access with bounded maximum delay.

Most of these works, however, addressed generic data communication applications. For more specific applications, such as real-time multimedia, it is possible to exploit the peculiar characteristics of the data, and in particular, their non-uniform importance, to further optimize the communication performance.

In this chapter a perceptually optimized video transmission algorithm aimed at providing reliable low-delay intervehicle video communications is presented. The key idea of the algorithm is to give more transmission priority to the most important packets, i.e., the ones which cause the highest distortion in the decoded video in case of loss. The algorithm is tested using actual intervehicle packet transmission traces and video sequences captured from on-board cameras. The performance of the algorithm is compared with two reference techniques, that is the standard 802.11 MAC-layer retransmission scheme and a delay-constrained retransmission scheme, showing that the gains are significant and consistent in different communication conditions.

The chapter is organized as follows. Section 7.2 briefly introduces the H.264 video coding standard, the analysis-by-synthesis algorithm used to evaluate perceptual importance, and provides an overview of 802.11-based intervehicle video communication issues. In Section 7.3, the perceptually optimized scheduling technique as well as the two reference techniques are analyzed in detail. The experimental setup is described in Section 7.4, while Section 7.5 provides extensive results. Finally, conclusions are drawn in Section 7.6.

7.2 The Inter-Vehicle Video Communications Scenario

7.2.1 The H.264 Video Coding Standard

This chapter focuses on intervehicle video communications based on the state-of-the-art H.264 video codec [6, 7], which is particularly suitable for packet networks communications. In fact, one of the most interesting characteristics of the H.264 standard is the decoupling of the coding aspects from the bitstream adaptation needed for transmission over a particular channel.

As in previous video coding standards, the H.264 video coding layer groups consecutive macroblocks into slices, that are the smallest independently decodable units. Slices are important because they allow to subdivide the coded bitstream into independent packets, so that the loss of a packet does not affect the ability of the receiver to decode the bitstream of others.

Different from other video coding standards, the H.264 provides a network abstraction layer which aims to efficiently support transmission over IP networks. In particular, it relies on the use of the real-time transport protocol (RTP), which is well suited for real-time wired and wireless multimedia transmissions, and it has been employed in the experiments presented in this chapter. However, some dependencies exist between the video coding and the network abstraction layer. For instance, the packetization process is improved if slice size is about the same of packet size and if only one slice is inserted in each packet, thus creating independently decodable packets. The packetization strategy, as the frame subdivision into slices, is not standardized and the encoder has the possibility to vary both of them for each frame. Usually, however, the maximum packet size (hence slice size) is limited and slices cannot be too short due to the resulting overhead that would reduce coding efficiency. Thus in the experiments one slice per packet is employed and the slice size is limited as described in Section 7.4.

7.2.2 Analysis-by-Synthesis Distortion Estimation for Video Packets

The quality of multimedia communications over packet networks may be impaired in case of packet loss. The amount of quality degradation strongly vary depending on the importance of the lost data. In order to design efficient loss protection mechanisms, a reliable importance estimation method for multimedia data is needed, so that the most important packets can be privileged by the transmission policy. The transmission policy optimization problem for multimedia traffic has been investigated in details in [8], which however relies on the availability of an importance value for each packet. Such importance is often defined a priori, based on the average importance of the elements of the compressed bitstream, as with the data-partitioning approach.

In order to provide a quantitative importance estimation method at a finer level of granularity, the algorithm proposed originally in [9] is employed. Thus the importance of each packet in terms of the distortion that would be introduced at the decoder by the loss of that specific packet. The value is indeed the mean squared error (MSE) between the original and the reconstructed picture after concealment and it will be referred to as the *distortion of the packet*. The analysis-by-synthesis algorithm is briefly summarized in the following. The algorithm performs, for each packet, the following steps:

1. Decoding, including concealment, of the bitstream simulating the loss of the packet being analyzed (synthesis stage)
2. Quality evaluation, that is, computation of the distortion caused by the loss of the packet. The original and the reconstructed picture after concealment are compared using, e.g., mean squared error (MSE)
3. Storage of the obtained value as an indication of the perceptual importance of the analyzed video packet

Previous operations can be implemented with small modifications of the standard encoding process. The encoder, in fact, reconstructs the coded pictures simulating the decoder operations, since this is needed for motion-compensated prediction. If step 1 of the analysis-by-synthesis algorithm exploits the operations of the encoding software, complexity is only due to the simulation of the concealment algorithm. In case of simple temporal concealment techniques, this is trivial and the task is reduced to provide the data to the quality evaluation algorithm.

The analysis-by-synthesis technique, as a principle, can be applied to any video coding standard. In fact, it is based on repeating the same steps that a standard decoder would perform, including error concealment. However, due to the interdependencies usually present between data units, the simulation of the loss of an isolated data unit is not completely realistic, particularly for high packet loss rates. Every possible combination of events should ideally be considered, weighted by its probability, and its distortion computed by the analysis-by-synthesis technique, obtaining the expected distortion value. For simplicity, however, we assume that all preceding data units have been correctly received and decoded. Nevertheless, this leads to a useful approximation as demonstrated by some applications of the analysis-by-synthesis approach to MPEG-coded video [9]. The results section will show the effectiveness of the video transmission algorithm which relies on these distortion values.

To reduce complexity, statistical studies on many different video sequences have been conducted and a model-based approach [10] has been developed.

According to that model the encoder computes the distortion that would be caused by the loss of the packet into the current frame and then, using a simple formula, it computes an estimation of the total distortion which includes future frames. It is worth noting that the use of the model extends the applicability of the analysis-by-synthesis distortion estimation algorithm to the case of a live encoding scenario, which is the focus of this chapter.

7.2.3 Multimedia Communications over 802.11

This chapter focuses on the well-known 802.11 [11] standard as the underlying protocol for intervehicle multimedia communications. As pointed out in the introduction, an extension (802.11p) is currently being developed to provide networking services and protocols in the WAVE communications standards.

Due to the intrinsic unreliability of wireless channels the original 802.11 standard implements an immediate per-packet acknowledgment scheme, which sends a positive acknowledgment MAC packet for each correctly received one. Packets which are not correctly received are retransmitted by the sender device up to a maximum number of retransmissions, which is referred to as retry limit. Hence in a point-to-point communication the MAC layer of each device is able to immediately know the outcome of its transmissions. However, traditional MAC implementations do not pass the information to higher layers in the protocol stack, hence higher layers are not notified in case of transmission failure. It is up to higher layer protocols, such as TCP, to detect packet losses and retransmit packets if needed.

In low-delay applications, such as audio and video conferencing, due to the maximum end-to-end delay allowed (150 ms [12]), end-to-end application-layer retransmission mechanisms are often too slow or would involve sending excessively frequent acknowledgment information in order to quickly retransmit corrupted or missing packets. Note also that the 802.11 access scheme is subject to a drastic performance decrease if the number of packets offered to the network exceeds a given threshold. For these reasons, using the MAC level acknowledgment information, which is available at no cost, is desirable in case a direct communication is performed (e.g., in 802.11 ad hoc mode). However, this requires a cross-layer communication between the MAC layer and the application layer to signal the outcome of the transmission of each MAC packet. Current 802.11 MAC implementations are designed to automatically retransmit the same packet a number of times up to the retry limit value, but they could be easily modified to communicate with the application layer to implement different retransmission strategies. Note that these modifications are fully compliant with the 802.11 standard since no changes in the receivers are needed.

This chapter focuses on improving the MAC retransmission mechanism of the 802.11 standard since retransmission techniques are well suited for the rapidly changing characteristics of intervehicle wireless channels. In fact, retransmission mechanisms automatically adapt to different channel conditions, they do not waste bandwidth when channel is good, and modifications are 802.11 standard compliant: only a slight modification of the software driver of the devices is required. Alternative solutions such as redundant codes, instead, have the main drawback of requiring additional hardware and memory, they are not standard compliant and important modifications of the MAC layer of devices are often needed. Finally, note that real-time multimedia data exhibit strong time sensitiveness. Packets which are not successfully received before their playback time should not be retransmitted since this would only result in bandwidth waste. Hence cross-layer communications between application and MAC layers is desirable because it allows to eliminate from the transmission queue the packets which cannot arrive on time at the decoder, thus reducing the delay of all the remaining queued packets.

7.3 The Perceptually Optimized Packet Scheduling Algorithm

This section presents a perceptually optimized packet scheduling algorithm for robust video communications over 802.11 network. To assess the performance gain, a comparison with two reference 802.11 packet scheduling algorithms is presented. The perceptually optimized algorithm takes advantage of the packet importance estimation technique as described in Section 7.2.2. That technique allows to compute, for each packet, an estimate of the distortion impact in case the packet is lost. Under the assumption that the distortion D of a sequence at the receiver can be computed, as a first approximation, as the sum of the distortion associated with the incorrectly received packets, the D value can be minimized giving priority, at each transmission opportunity, to the packet which presents the highest distortion value. Thus, each time a new packet can be transmitted the one with the highest distortion value is selected and sent. The validity of this assumption is confirmed by the simulation results.

Note that the sender has perfect knowledge of the status of the communication by means of the MAC acknowledgment packets, i.e., it knows which packets have been received and which have not, if the lost acknowledgment case is neglected. Moreover, packets which cannot reach the receiving node on time for playback are immediately dropped from the transmission queue not to waste transmission resources. Note also that there are no a priori limits on the number of times a packet can be retransmitted, differently from the other two techniques. However, the average number of retransmissions is similar for all the algorithms. The algorithm will be referred to as perceptually optimized scheduling (POS) in the rest of the chapter.

To assess the impact of the POS algorithm, first we compared it with the standard MAC-layer ARQ scheme, which retransmits non-acknowledged packets up to a certain number of times, given by the retry limit value. This algorithm is implemented in all currently available 802.11 devices. Various retry limit values have been used in the comparisons. Note also that the standard MAC-layer ARQ scheme discards packets which have been queued at the transmission interface for more than a certain time (referred to as lifetime in the standard, default value is about 500 ms).

The second reference algorithm uses the same retransmission policy of the standard MAC-layer ARQ algorithm, but it immediately eliminates packets from the transmission queue when it determines that they cannot reach the receiving node on time for playback. This requires a form of communication between the application layer, which knows the deadline of each packet, and the MAC layer, which discards packets. Due to this behavior the algorithm is referred to as "maximum-delay ARQ" (MD-ARQ) in the rest of the chapter.

7.4 Experimental Setup

An 802.11 transmission between two vehicles traveling along the same road has been simulated by means of packet level error traces obtained with intervehicle transmission experiments [13, 14]. In these experiments, packets of 1470 bytes were continuously transmitted between two vehicles using 802.11 devices operating at physical transmission speed of 2 Mbit/s, reaching an approximate packet rate of 150 packets per second. An external antenna was used to improve signal reception. Each trace has been collected trying to keep the distance from the preceding car as constant as possible, but variations are present due to traffic conditions.

Two different application scenarios, which however present similar constraints, have been considered. In the first one, road video is captured from a front on-board camera, then compressed with the state-of-the-art H.264 video compression standard and sent to the second vehicle, which can use it, for instance, for cooperative visual processing of road information. In the second scenario, a video of the driver's face including the inside of the vehicle is captured by an on-board camera and sent to the other vehicle, thus simulating, for instance, a videoconference application between passengers in the two cars.

All videos were captured at a resolution of 640×480 pixels, at 30 fps, compressed at high bitrate and stored. They were later re-encoded using the H.264 codec [15], simulating a live encoding with a fixed quantization stepsize, so that video quality is kept approximately constant. The simulations assume the use of the IP/UDP/RTP protocol stack which is well suited for real-time multimedia transmissions. All simulations use a fixed pattern of frame types, i.e., one I-type frame followed by 11 P-type frames. The length of the video segments used in the experiments is 30 s, and all results are averaged over four simulations using consecutive segments of a given packet loss trace.

Video quality has been evaluated using the peak signal-to-noise ratio (PSNR) measure which, taking in due consideration its well-known limits, is widely accepted in the multimedia communications research community. The encoding video quality is 38.83 and 44.16 dB for the "road video" and "driver's face" sequences, respectively. In case of missing packets at the decoder, a simple temporal concealment technique is implemented, that is the missing macroblocks are replaced by the data in the same area in the previous frame. Finally, for simulation simplicity, acknowledgment MAC packets are always assumed to be correctly received, for all the three techniques.

7.5 Results

This section analyzes the performance of the POS algorithm compared with the two reference ARQ techniques, that is, the standard MAC-layer ARQ and the MD-ARQ techniques. The characteristics of the packet loss traces used in the experiments are shown in Table 7.1. However, due to the variability of the traffic conditions, each single segment of the traces may present values which differ from the averages shown in the table. Note also that, besides the distance between the two cars, other factors may affect channel quality, i.e., vehicular traffic conditions and presence of obstacles.

Figure 7.1 shows a sample segments of the packet loss traces used in the experiments. Each black square represents a packet transmission error, while white color indicates correct packet reception. Each row represents about 1.33 s of transmission. Depending on the time instant, errors may be sporadic or concentrated in bursts, whose duration is proportional to the length of the black rectangles. Note also that wireless channel conditions change very rapidly from near-perfect to very impaired communication. This confirms that packet retransmission mechanisms are suitable for intervehicle wireless channels.

Figure 7.2, which refers to the case of transmission of road video, shows the video quality performance of the POS algorithm, compared to the other two

Table 7.1 Characteristics of the packet loss traces

Trace	Avg. packet loss rate (%)	Avg. car-to-car distance (m)
A	9.6	320
B	27.5	360
C	28.0	398
D	36.7	415

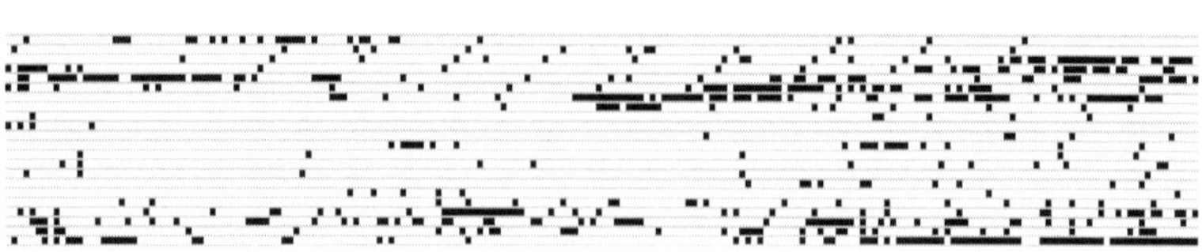

Fig. 7.1 Samples of error traces used in the experiments. Each *black square* represents a packet transmission error. Figure shows the fourth segment of trace B

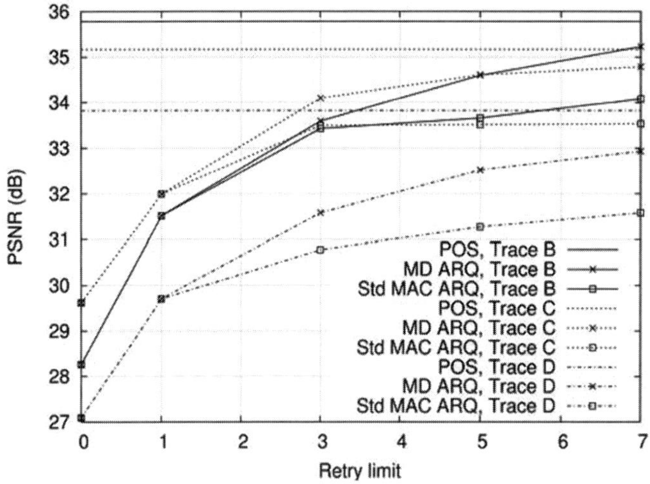

Fig. 7.2 PSNR performance of the proposed POS technique compared with the two reference techniques. Maximum delay set to 150 ms, sequence: "road video"

reference algorithms. Different values of the retry limit parameters have been tested for the standard MAC-layer ARQ and the MD-ARQ algorithms. The performance of the POS algorithm is shown for comparison as horizontal lines. The video quality increases as a function of the retry limit for both the standard MAC-layer ARQ and the MD-ARQ algorithms. When they achieve the maximum performance, i.e., the retry limit is set to seven, the POS technique presents a gain up to 1 dB over the MD-ARQ technique, and up to 2.2 dB when compared with the standard MAC-layer ARQ technique. Moreover, note that simply discarding packets which cannot arrive on time at the receiver (MD-ARQ algorithm) greatly improves the quality performance over the standard MAC-layer ARQ, more than 1 dB when the retry limit is equal to seven.

Figure 7.3 shows the results of the same set of experiments for the case of videoconferencing, i.e., the camera is aimed at the driver's face and inside the vehicle. Performance trends and gains are similar to the case of road video. The absolute PSNR values are different since the encoding distortion of the two video sequences is different.

The amount of packet losses as seen from the application layer is shown in Fig. 7.4. When the retry limit is increased, the packet loss rate of the standard MAC-layer ARQ and the MD-ARQ techniques decreases. However, while the trend is clear for the MD-ARQ technique, the packet loss rate of the standard MAC-layer ARQ technique presents a minimum, that is achieved with retry limit equal to three in Fig. 7.4. This is because the increased number of retransmission attempts cause packets to accumulate in the device queue in case of bad channel conditions. Note that, in any case, when the end-to-end delay of packets is higher than the maximum allowed value, that is fixed to

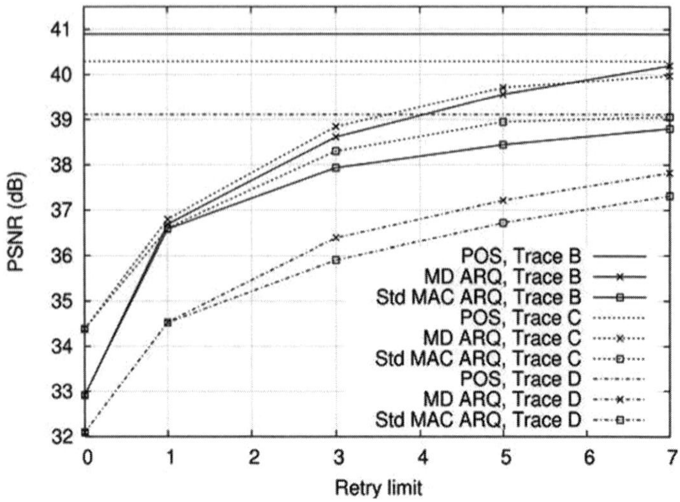

Fig. 7.3 PSNR performance of the proposed POS technique compared with the two reference techniques. Maximum delay set to 150 ms, sequence: "driver's face"

150 ms in our simulations, packets are discarded at the receiver because they are too late for playback. But for the case of the standard MAC-layer ARQ, the sending device drops packets from the transmission queue only when the time spent in the queue is greater than the lifetime (about 500 ms). This results in an

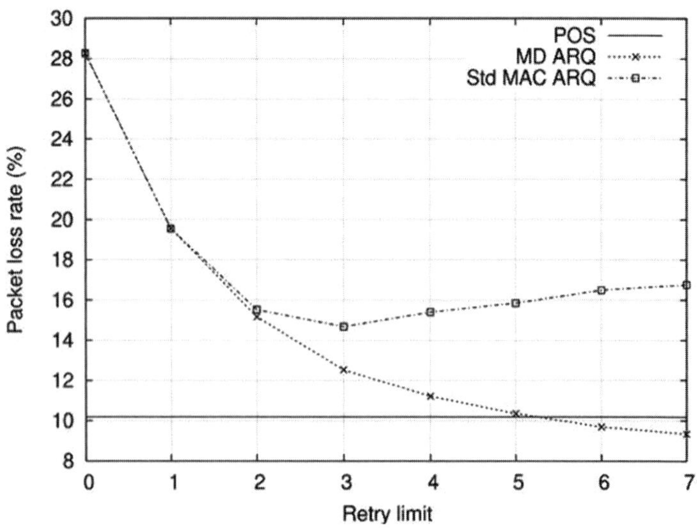

Fig. 7.4 Packet loss rate as seen from the application layer. Trace B, maximum delay set to 150 ms, sequence: "road video"

Table 7.2 Average number of MAC-layer retransmissions; maximum delay set to 150 ms, retry limit equals to seven, sequence: "road video"

Trace	Std MAC ARQ	MD-ARQ	POS
B	1.62	1.58	1.57
C	1.55	1.51	1.50
D	1.74	1.70	1.69

increase of the delay of queued packets. This effect has been investigated in details in [16]. Packets transmitted using the MD-ARQ algorithm, instead, do not experience the delay increase because they are immediately dropped from the transmission queue if they wait for more than the maximum allowed delay, i.e., 150 ms.

Note also that, despite the MD-ARQ technique presents a lower packet loss rate compared to the POS technique, the PSNR performance of the latter is higher. The slightly higher number of losses is, indeed, counterbalanced by the fact that they systematically affect the less important packets.

Table 7.2 shows the average number of transmissions of MAC-layer packets, for the same experiments of Fig. 7.4. The standard MAC-layer ARQ technique presents the highest average number of retransmissions but, as previously explained, many packets are then dropped by the application layer of the receiver because they are too late for playback. The performance of the MD-ARQ and the POS technique is similar for high retry limit values. However, note that the POS technique presents a slightly lower channel usage in terms of number of transmitted MAC packets, while the PSNR value is higher than the other techniques, as shown in Fig. 7.2.

The performance of the POS and the two reference ARQ techniques has also been assessed as a function of the maximum allowed delay. Results are shown in Fig. 7.5. The POS technique shows a clear PSNR performance increase (up to 0.7 dB) compared with the MD-ARQ technique when the maximum allowed delay is increased over 100 ms, which is the time interval corresponding to three video frames. The performance of the standard MAC-layer ARQ technique is much lower, about 1 dB less than the MD-ARQ technique, regardless of the maximum allowed delay, because it is greatly influenced by packet accumulation in the queue when channel conditions are bad. For the less interesting case of low retry limit values (e.g., three), the PSNR does not increase significantly for both the standard MAC-layer ARQ and the MD-ARQ when the maximum delay is increased.

Summarizing previous results, the best conditions for all the techniques are given by the maximum delay set to 150 ms and the retry limit fixed to seven. In this case, the video quality values confirm the consistency of the performance gain achieved by the POS technique. In particular, the gain is up to about 2.6 dB with respect to the standard MAC-level ARQ technique, and up to about 1.3 dB when considering the MD-ARQ technique, depending on the trace and video

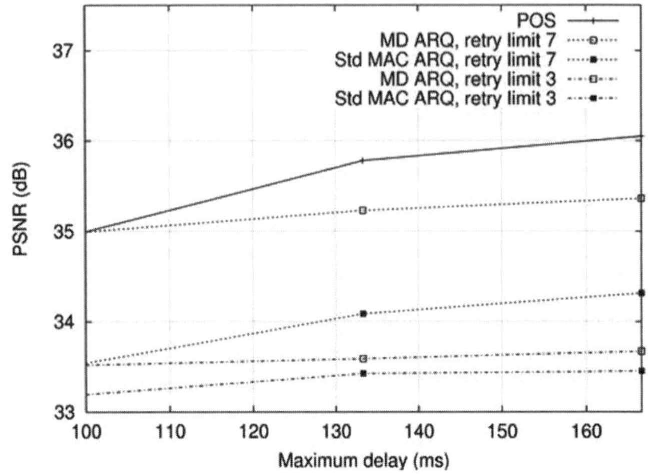

Fig. 7.5 PSNR performance as a function of the maximum allowed delay. Trace B, sequence: "road video"

Fig. 7.6 (a) Visual performance of the standard MAC-layer ARQ technique. **(b)** Visual performance of the MD-ARQ technique. **(c)** Visual performance of the POS technique

sequence. For the case of Trace A (not shown in graphs), the performance gap is limited since most packet losses can be recovered, however, values are close to their maximum, that is the quality achieved by the H.264-encoding process.

Finally, a visual comparison is provided. Figure 7.6 shows the same decoded video frame for the three ARQ techniques; maximum delay is set to 150 ms, retry limit is equal to seven. The distortion, which is evident in the case of the MAC-layer ARQ technique, is reduced but still noticeable on the side of the road and on the guard rail with the MD-ARQ technique. The picture provided by the POS technique is still affected by some distortion but the road side and the guard rail are better than the case of the MD-ARQ technique.

7.6 Conclusions

This chapter presented a perceptually optimized scheduling algorithm which can achieve robust low-delay transmission of video data captured by an on-board camera to another vehicle in proximity using the 802.11 wireless technology. The algorithm is based on exploiting the non-uniform importance of packets, defined as the distortion which they may cause at the decoder in case of loss. Transmission of in-car and road video has been simulated using packet level error traces collected in vehicle-to-vehicle communication experiments. The results show that the perceptually optimized packet scheduling algorithm achieves a consistently higher quality video communication compared to two reference techniques, with gains up to about 2 dB PSNR. The consistency of the improvements has been shown using different error traces and various maximum-delay settings. Moreover, the results also show that to achieve an efficient low-delay communication in high loss rate scenarios such as intervehicle communications it is very important to eliminate packets from the 802.11 retransmission queue as soon as they are no more useful for the application, not to impact on the end-to-end transmission delay of all the remaining packets.

Acknowledgments The authors would like to thank in particular Professor K. Takeda, Professor H. Abut, and all the participants of the NEDO project [17], who collected and provided the video traces used in this chapter, and the CRAWDAD project [13] for the intervehicle transmission traces used in these experiments.

References

1. Ott J, Kutscher D (2004) Drive-thru Internet: IEEE 802.11b for automobile users. Proceedings of the IEEE INFOCOM 1:362–373.
2. Singh JP, Bambos N, Srinivasan B et al. (2002) Wireless LAN performance under varied stress conditions in vehicular traffic scenarios. Proceedings of the IEEE Vehicular Technology Conference (VTC) 2:743–747.

3. Wang SY, Lin CC, Hwang YW et al. (2005) A practical routing protocol for vehicle-formed mobile ad hoc networks on the roads. Proceedings of IEEE International Conference on Intelligent Transportation Systems 161–166
4. Wischhof L, Ebner A, Rohling H (2005) Information dissemination in self-organizing intervehicle networks. IEEE Transactions on Intelligent Transportation Systems 6(1):90–101
5. Pal A, Dogan A, Ozguner F et al. (2002) A MAC layer protocol for real-time intervehicle communication. Proceedings of IEEE International Conference on Intelligent Transportation Systems 353–358
6. ITU-T (2003) Advanced video coding for generic audiovisual services. Rec H.264 & ISO/IEC 14496-10 AVC.
7. Wiegand T, Sullivan GJ, Bjontegaard G et al. (2003) Overview of the H.264/AVC video coding standard. IEEE Transactions on Circuits and Systems for Video Technology 13(7): 560–576
8. Chou PA, Miao Z (2006) Rate-distortion optimized streaming of packetized media. IEEE Transactions on Multimedia 8(2):390–404
9. Masala E, De Martin JC (2003) Analysis-by-synthesis distortion computation for rate-distortion optimized multimedia streaming. Proceedings of the IEEE International Conference on Multimedia & Expo 3:345–348
10. De Vito F, Quaglia D, De Martin JC (2004) Model-based distortion estimation for perceptual classification of video packets. Proceedings of the IEEE International Workshop on Multimedia Signal Processing (MMSP) 1:79–82
11. ISO/IEC (1999) Wireless LAN Medium Access Control (MAC) and Physical Layer (PHY) Specifications. ISO/IEC 8802-11, ANSI/IEEE Std 802.11
12. ITU-T (2002) One-way transmission time. Rec G.114
13. CRAWDAD Project (2007) A community resource for archiving wireless data at Dartmouth. crawdad.cs.dartmouth.edu. Accessed April 2007
14. Wu H, Palekar M, Fujimoto R et al. (2005) An empirical study of short range communications for vehicles. Proceedings of the 2nd ACM International Workshop on Vehicular Ad Hoc Networks 83–84.
15. Joint Video Team (JVT) of ISO/IEC MPEG & ITU-T VCEG (2007) Joint Model v.11.0. iphome.hhi.de/suehring/tml. Accessed January 2007
16. Li Q, Van Der Schaar M (2004) Providing adaptive QoS to layered video over wireless local area networks through real-time retry limit adaptation. IEEE Transactions on Multimedia 6(2):278–290
17. NEDO Project, Driving behavior signal processing based on large scale real world database. www.sp.m.is.nagoya-u.ac.jp/NEDO.

8
Machine Learning Systems for Detecting Driver Drowsiness

Esra Vural, Müjdat Çetin, Aytül Erçil, Gwen Littlewort, Marian Bartlett, and Javier Movellan

Abstract Drowsy driver detection is one of the potential applications of intelligent vehicle systems. Previous approaches to drowsiness detection primarily make pre-assumptions about the relevant behavior, focusing on blink rate, eye closure, and yawning. Here we employ machine learning to datamine actual human behavior during drowsiness episodes. Automatic classifiers for 30 facial actions from the facial action coding system were developed using machine learning on a separate database of spontaneous expressions. These facial actions include blinking and yawn motions, as well as a number of other facial movements. These measures were passed to learning-based classifiers such as Adaboost and multinomial ridge regression. Head motion information was collected through automatic eye tracking and an accelerometer. The system was able to predict sleep and crash episodes on a simulator with 98% accuracy across subjects. It is the highest prediction rate reported to date for detecting drowsiness. Moreover, the analysis revealed new information about human facial behavior for drowsy drivers.

Keywords Driver fatigue · Drowsiness · Machine learning · Facial expressions · Facial action unit · Head movements · Multinomial logistic regression · Support vector machine (SVM) · Coupling · Driver behavior

8.1 Introduction

In recent years, there has been growing interest in intelligent vehicles. A notable initiative on intelligent vehicles was created by the US Department of Transportation with the mission of prevention of highway crashes [1]. The on-going

E. Vural (✉)
Faculty of Engineering and Natural Sciences, Sabancı University, Orhanli, Istanbul, Turkey
e-mail: esravural@su.sabanciuniv.edu

K. Takeda et al. (eds.), *In-Vehicle Corpus and Signal Processing for Driver Behavior*, DOI 10.1007/978-0-387-79582-9_8,

intelligent vehicle research is expected to revolutionize the way vehicles and drivers interact in the future.

The US National Highway Traffic Safety Administration estimates that in the USA alone approximately 100,000 crashes each year are caused primarily by driver drowsiness or fatigue [2]. Thus incorporating automatic driver fatigue detection mechanisms into vehicles may help prevent many accidents.

One can use a number of different techniques for analyzing driver exhaustion. One set of techniques places sensors on standard vehicle components, e.g., steering wheel, gas pedal, and analyzes these sensor signals to detect drowsiness [3]. Such techniques may need to be adapted to the driver, since there are noticeable differences among drivers in the way they use the gas pedal [4].

A second set of techniques focuses on measurement of physiological signals such as heart rate, pulse rate, and electroencephalography (EEG) [5]. It has been reported by researchers that as the alertness level decreases, power of the alpha and theta bands in the EEG signal increases [6], providing indicators of drowsiness. However, this method has drawbacks in terms of practicality since it requires a person to wear an EEG cap while driving.

A third set of solutions focuses on computer vision systems that can detect and recognize the facial motion and appearance changes occurring during drowsiness [7, 8]. The advantage of computer vision techniques is that they are non-invasive, and thus are more amenable to use by the general public. There are some significant previous studies about drowsiness detection using computer vision techniques. Most of the published research on computer vision approaches to detection of fatigue has focused on the analysis of blinks and head movements. However, the effect of drowsiness on other facial expressions has not been studied thoroughly. Recently Gu and Ji presented one of the first fatigue studies that incorporates certain facial expressions other than blinks [9]. Their study feeds action unit information as an input to a dynamic Bayesian network. The network was trained on subjects posing a state of fatigue. The video segments were classified into three stages: inattention, yawn, or falling asleep. For predicting falling asleep, head nods, blinks, nose wrinkles, and eyelid tighteners were used.

Earlier approaches to drowsiness detection primarily make pre-assumptions about the relevant behavior, focusing on blink rate, eye closure, and yawning. Here, we use machine learning methods to datamine the human behavior during drowsiness episodes. The goal of this study is to study what facial configurations are predictors of fatigue. In other words, our aim is to discover the significant associations between a single or a combination of facial expressions and fatigue. In this project, facial motion was analyzed automatically from video using a fully automated facial expression analysis system based on the facial action coding system (FACS) [10]. In addition to the output of the automatic FACS recognition system, we have also collected head motion data using an accelerometer placed on the subject's head, as well as steering wheel data. Adaboost and Multinomial logistic regression-based classifiers have been employed while discovering the highly significant facial expressions and detecting fatigue.

8.2 Methods

8.2.1 Driving Task

Subjects played a driving video game on a windows machine using a steering wheel[1] and an open source multi-platform video game[2] (see Fig. 8.1). The windows version of the video game was maintained such that at random times, a wind effect was applied that dragged the car to the right or left, forcing the subject to correct the position of the car. This type of manipulation had been found in the past to increase fatigue [11]. Driving speed was held constant. Four subjects performed the driving task over a 3-h period beginning at midnight. During this time subjects fell asleep multiple times thus crashing their vehicles. Episodes in which the car left the road (crash) were recorded. Video of the subjects face was recorded using a digital video camera for the entire 3-h session.

Figure 8.2 shows an example of a subject falling asleep during this task. First we see the eyes closing and drifting away from the centerline, followed by an overcorrection. Then the eyes close again, there is more drift, followed by a crash. We investigated facial behaviors that predicted these episodes of falling asleep and crashing and 60 s before crashes are taken as drowsy episodes. Given

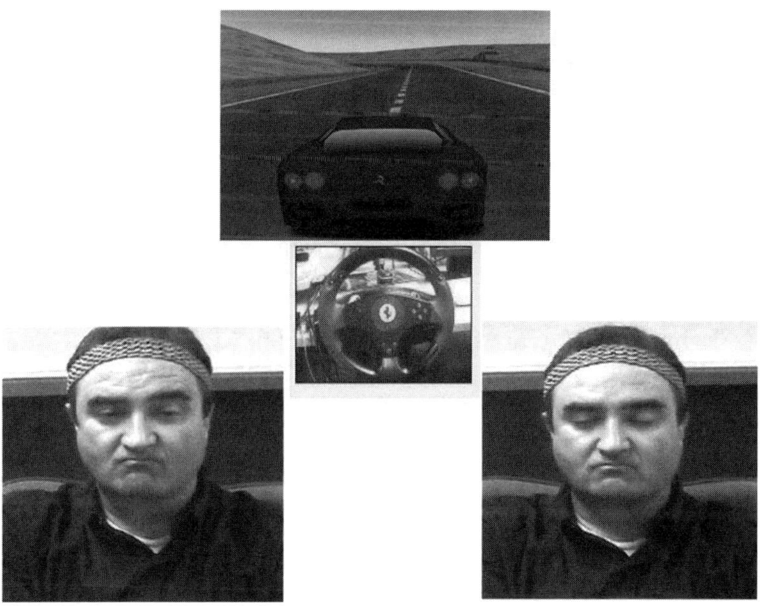

Fig. 8.1 Driving simulation task

---

[1] Thrustmaster® Ferrari Racing Wheel.

[2] Open Racing Car Simulator (TORCS).

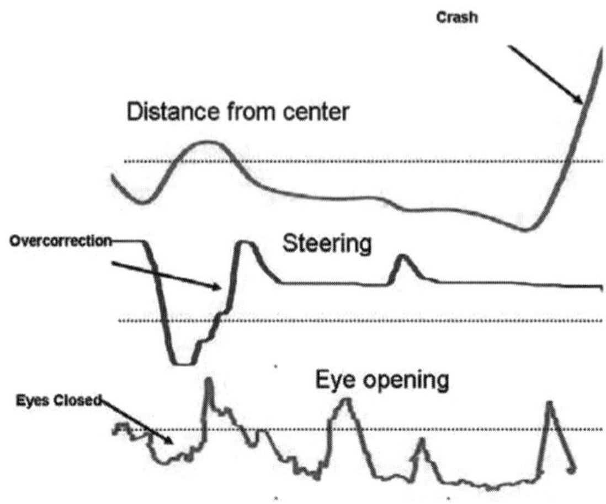

Fig. 8.2 Driving signals and eye opening signal (eye opening signal is obtained by inverting the eye closure signal (action unit 45). See Section 8.2.3 for a list of facial action coding signals and a description of the action unit)

a video segment the machine learning task is to predict whether the segment is coming from an alert or drowsy episode.

8.2.2 Head Movement Measures

Head movement was measured using an accelerometer that has 3 degrees of freedom. This three-dimensional accelerometer[3] has three one-dimensional accelerometers mounted at right angles measuring accelerations in the range of $-5g$ to $+5g$ where g represents earth's gravitational force.

8.2.3 Facial Action Classifiers

The facial action coding system (FACS) [12] is arguably the most widely used method for coding facial expressions in the behavioral sciences. The system describes facial expressions in terms of 46 component movements, which roughly correspond to the individual facial muscle movements. An example is shown in Fig. 8.3. FACS provides an objective and comprehensive way to analyze expressions into elementary components, analogous to decomposition of speech into phonemes. Because it is comprehensive, FACS has proven useful for discovering facial movements that are indicative of cognitive and affective states. In this chapter we investigate whether there are action units (AUs) such as chin raises (AU17), nasolabial furrow deepeners (AU11), outer (AU2) and

[3] Vernier ®.

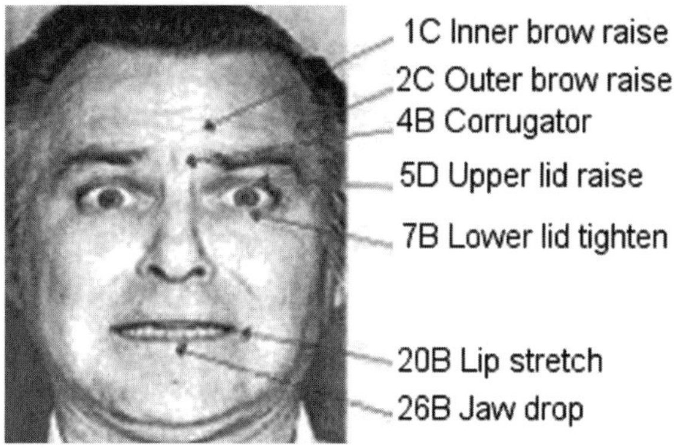

Fig. 8.3 Sample of facial action decomposition from the facial action coding system [12]

inner brow raises (AU1) that are predictive of the levels of drowsiness observed prior to the subjects falling asleep.

In previous work we have presented a novel system, named CERT, for fully automated detection of facial actions from the facial action coding system [10]. The workflow of the system is summarized in Fig. 8.4. We previously reported detection of 20 facial action units, with a mean of 93% correct detection under controlled posed conditions and 75% correct for less controlled spontaneous expressions with head movements and speech.

For this project we used an improved version of CERT which was retrained on a larger data set of spontaneous as well as posed examples. In addition, the system was trained to detect an additional 11 facial actions for a total of 31 (see Table 8.1). The facial action set includes blink (action unit 45), as well as facial actions involved in yawning (action units 26 and 27). The selection of this set of 31 out of 46 total facial actions was based on the availability of labeled training data.

For this study we have employed support vector machines. One support vector machine (SVM) was trained for each of the 31 facial actions, and it was trained to detect the facial action regardless of whether it occurred alone or in

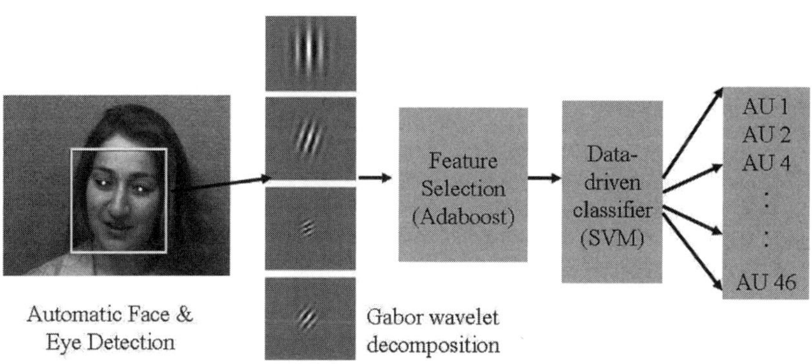

Fig. 8.4 Overview of the proposed fully automated facial action coding system

Table 8.1 Full set of action units used for predicting drowsiness

AU	Name	AU	Name
1	Inner brow raise	18	Lip pucker
2	Outer brow raise	19	Tongue show
4	Brow lowerer	20	Lip stretch
5	Upper lid raise	22	Lip funneler
6	Cheek raise	23	Lip tightener
7	Lids tight	24	Lip presser
8	Lip toward	25	Lips part
9	Nose wrinkle	26	Jaw drop
10	Upper lip raiser	27	Mouth stretch
11	Nasolabial furrow deepener	28	Lips suck
12	Lip corner puller	30	Jaw sideways
13	Sharp lip puller	32	Bite
14	Dimpler	38	Nostril dilate
15	Lip corner depressor	39	Nostril compress
16	Lower lip depress	45	Blink
17	Chin raise		

combination with other facial actions. The system output consists of a continuous value which is the distance to the separating hyperplane for each test frame of video. The system operates at about 6 frames/s on a Mac G5 dual processor with 2.5 GHz processing speed.

Facial expression training data: The training data for the facial action classifiers came from two posed data sets and one data set of spontaneous expressions. The facial expressions in each data set were FACS coded by certified FACS coders. The first posed data sets was the Cohn-Kanade DFAT-504 data set [13]. This data set consists of 100 university students who were instructed by an experimenter to perform a series of 23 facial displays, including expressions of seven basic emotions. The second posed data set consisted of directed facial actions from 24 subjects collected by Ekman and Hager [12]. Subjects were instructed by a FACS expert on the display of individual facial actions and action combinations, and they practiced with a mirror. The resulting video was verified for AU content by two certified FACS coders. The spontaneous expression data set consisted of a set of 33 subjects collected by Mark Frank at Rutgers University. These subjects underwent an interview about political opinions on which they felt strongly. Two minutes of each subject were FACS coded. The total training set consisted of 6000 examples, 2000 from posed databases and 4000 from the spontaneous set.

8.3 Results

Subject data was partitioned into drowsy (non-alert) and alert states as follows. The 1 min preceding a sleep episode or a crash was identified as a non-alert state for each of the four subjects. There was an intersubject mean of 24 non-alert episodes with a minimum of 9 and a maximum of 35. Fourteen alert segments for each subject were collected from the first 20 min of the driving task. Our initial analysis focused on drowsiness prediction within-subjects.

8.3.1 Facial Action Signals

The output of the facial action detector consisted of a continuous value for each frame which was the distance to the separating hyperplane, i.e., the margin. Histograms for two of the action units in alert and non-alert states are shown in Fig. 8.5. The area under the ROC (A') was computed for the outputs of each facial action detector to see to what degree the alert and non-alert output distributions were separated. The A' measure is derived from signal detection theory and characterizes the discriminative capacity of the signal, independent of decision threshold. A' can be interpreted as equivalent to the theoretical maximum percent correct achievable with the information provided by the system when using a 2-Alternative Forced Choice testing paradigm. Table 8.2 shows the actions with the highest A' for each subject. As expected, the blink/eye closure measure was overall the most discriminative for most subjects. However, note that for subject 2, the outer brow raise (action unit 2) was the most discriminative.

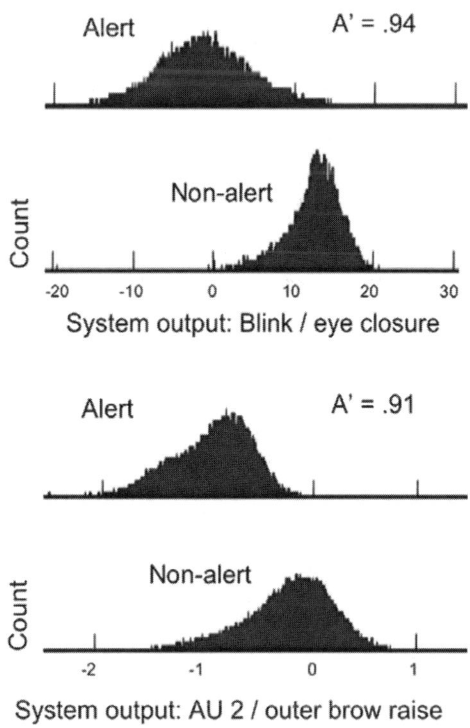

Fig. 8.5 Histograms for the blink and the action unit 2 in alert and non-alert states. Here A' represents the area under the ROC

Table 8.2 Top five highest discriminant action units for discriminating alert from non-alert states for each of the four subjects. A' is the area under the ROC curve

Subjects	AU	Name	A'
Subj1	45	Blink	0.94
	17	Chin raise	0.85
	30	Jaw sideways	0.84
	7	Lid tighten	0.81
	39	Nostril compress	0.79
Subj2	2	Outer brow raise	0.91
	45	Blink	0.80
	17	Chin raise	0.76
	15	Lip corner depress	0.76
	11	Nasolabial furrow	0.76
Subj3	45	Blink	0.86
	9	Nose wrinkle	0.78
	25	Lips part	0.78
	1	Inner brow raise	0.74
	20	Lip stretch	0.73
Subj4	45	Blink	0.90
	4	Brow lower	0.81
	15	Lip corner depress	0.81
	7	Lid tighten	0.80
	39	Nostril compress	0.74

8.3.2 Drowsiness Prediction

Facial action outputs were passed to a classifier for predicting drowsiness based on the automatically detected facial behavior. Two learning-based classifiers, Adaboost and multinomial ridge regression, are compared. Within-subject prediction of drowsiness and across-subject (subject-independent) prediction of drowsiness were both tested.

Within-Subject Drowsiness Prediction: For the within-subject prediction, 80% of the alert and non-alert episodes were used for training and the other 20% were reserved for testing. This resulted in a mean of 19 non-alert and 11 alert episodes for training, and 5 non-alert and 3 alert episodes for testing per subject (Table 8.3).

The weak learners for the Adaboost classifier consisted of each of the 31 facial action detectors. The classifier was trained to predict alert or non-alert from each frame of video. There was a mean of 54,000 training samples, $(19+11)\times60\times30$, and 14,400 testing samples, $(5 + 3) \times 60 \times 30$, for each subject. On each training iteration, Adaboost selected the facial action detector that minimized prediction error given the previously selected detectors. Adaboost obtained 92% correct accuracy for predicting driver drowsiness based on the facial behavior.

Table 8.3 Performance for drowsiness prediction, within subjects. Means and standard deviations are shown across subjects

Classifier	Percent correct	Hit rate	False alarm rate
Adaboost	0.92 ±0.03	0.92 ±0.01	0.06±0.1
MLR	0.94 ± 0.02	0.98 ± 0.02	0.13 ± 0.02

Classification with Adaboost was compared to that using multinomial logistic regression (MLR). Performance with MLR was similar, obtaining 94% correct prediction of drowsy states. The facial actions that were most highly weighted by MLR also tended to be the facial actions selected by Adaboost; 85% of the top ten facial actions as weighted by MLR were among the first 10 facial actions to be selected by Adaboost.

Across-Subject Drowsiness Prediction: The ability to predict drowsiness in novel subjects was tested by using a leave-one-out cross-validation procedure. The data for each subject were first normalized to zero-mean and unit standard deviation before training the classifier. MLR was trained to predict drowsiness from the AU outputs several ways.

Performance was evaluated in terms of area under the ROC. As a single frame may not provide sufficient information, for all of the novel subject analysis, the MLR output for each feature was summed over a temporal window of 12 s (360 frames) before computing A'. MLR trained on all features obtained an A' of 0.90 for predicting drowsiness in novel subjects.

Finally, a new MLR classifier was trained by sequential feature selection, starting with the most discriminative feature (AU45), and then iteratively adding the next most discriminative feature given the features already selected. These features are shown at the bottom of Table 8.4. Best performance of 0.98 was obtained with five features: 45, 2, 19 (tongue show), 26 (jaw drop), and 15. This five-feature model outperformed the MLR trained on all features.

Effect of Temporal Window Length: Next, we have examined the effect on performance of the temporal window size. The five-feature model was employed for this

Table 8.4 Drowsiness detection performance for across subjects, using an MLR classifier with different feature combinations. The weighted features are summed over 12 s before computing A'

Feature	A'
AU45	0.9468
AU45,AU2	0.9614
A45,AU2,AU19	0.9693
AU45,AU2,AU19,AU26	0.9776
AU45,AU2,AU19,AU26,AU15	0.9792
All features	0.8954

Table 8.5 MLR model for predicting drowsiness across subjects

More when critically drowsy			Less when critically drowsy		
AU	Name	A'	AU	Name	A'
45	Blink/eye closure	0.94	12	Smile	0.87
2	Outer brow raise	0.81	7	Lid tighten	0.86
15	Lip corner depressor	0.80	39	Nostril compress	0.79
17	Chin raiser	0.79	4	Brow lower	0.79
9	Nose wrinkle	0.78	26	Jaw drop	0.77
30	Jaw sideways	0.76	6	Cheek raise	0.73
20	Lip stretch	0.74			
11	Nasolabial furrow	0.71			
14	Dimpler	0.71			
1	Inner brow raise	0.68			
10	Upper lip raise	0.67			
27	Mouth stretch	0.66			
18	Lip pucker	0.66			
22	Lip funneler	0.64			
24	Lip presser	0.64			
19	Tongue show	0.61			

analysis. The performances shown in Table 8.5 are obtained using a temporal window of 12 s. Here, the MLR output in the five-feature model was summed over windows of N seconds, where N ranged from 0.5 to 60 s. Figure 8.6 shows the area under the ROC for drowsiness detection in novel subjects over time periods. Performance saturates at about 0.99 as the window size exceeds 30 s. In other words, given a 30-s video segment the system can discriminate sleepy versus non-sleepy segments with 0.99 accuracy across subjects.

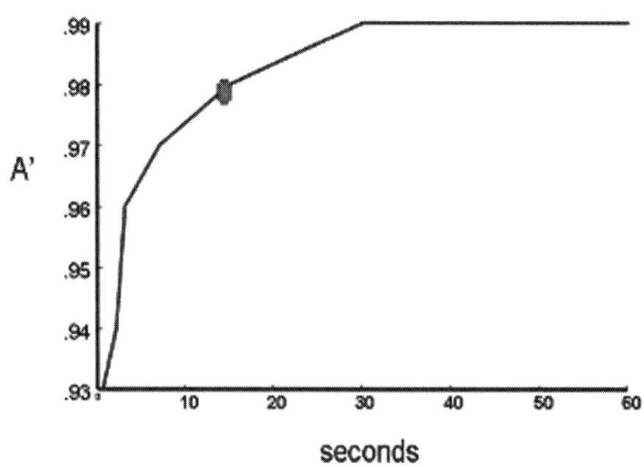

Fig. 8.6 Performance for drowsiness detection in novel subjects over temporal window sizes. *Large point* indicates the priorly obtained performance in Table 8.4 for a temporal window of 12 s using the five-feature model

Action Units Associated with Drowsiness: In order to understand how each action unit is associated with drowsiness, MLR was trained on each facial action individually. Examination of the A' for each action unit reveals the degree to which each facial movement was able to predict drowsiness in this study. The A's for the drowsy and alert states are shown in Table 8.5. The five facial actions that were the most predictive of drowsiness by increasing in drowsy states were 45, 2 (outer brow raise), 15 (frown), 17 (chin raise), and 9 (nose wrinkle). The five actions that were the most predictive of drowsiness by decreasing in drowsy states were 12 (smile), 7 (lid tighten), 39 (nostril compress), 4 (brow lower), and 26 (jaw drop). The high predictive ability of the blink/eye closure measure was expected. However, the predictability of the outer brow raise (AU2) was previously unknown.

It was observed during this study that many subjects raised their eyebrows in an attempt to keep their eyes open, and the strong association of the AU2 detector is consistent with that observation. Also of note is that action 26, jaw drop, which occurs during yawning, actually occurred less often in the critical 60 s prior to a crash. This is consistent with the prediction that yawning does not tend to occur in the final moments before falling asleep.

8.3.3 Coupling of Behaviors

As a preliminary work we analyzed the coupling between behaviors. Here you can find our preliminary results for coupling between first for steering and head motion and then for eye openness and eyebrow raises,

Coupling of Steering and Head Motion: Observation of the subjects during drowsy and alert states indicated that the subjects head motion differed substantially when alert versus when the driver was about to fall asleep. Surprisingly, head motion increased as the driver became drowsy, with large roll motion coupled with the steering motion as the driver became drowsy. Just before falling asleep, the head would become still.

We have also investigated the coupling of the head and arm motions. Correlations between head motion as measured by the roll dimension of the accelerometer output and the steering wheel motion are shown in Fig. 8.7. For this subject (subject 2), the correlation between head motion and steering increased from 0.27 in the alert state to 0.65 in the non-alert state. For subject 1, the correlation between head motion and steering similarly increased from 0.24 in the alert state to 0.43 in the non-alert state. The other two subjects showed a smaller coupling effect. Future work includes combining the head motion measures and steering correlations with the facial movement measures in the predictive model.

Coupling of Eye Openness and Eyebrow Raises: Our observations indicated that for some of the subjects coupling between eye brow ups and eye openness increased in the drowsy state. In other words subjects tried to open their eyes using their eyebrows in an attempt to keep awake (Fig. 8.8).

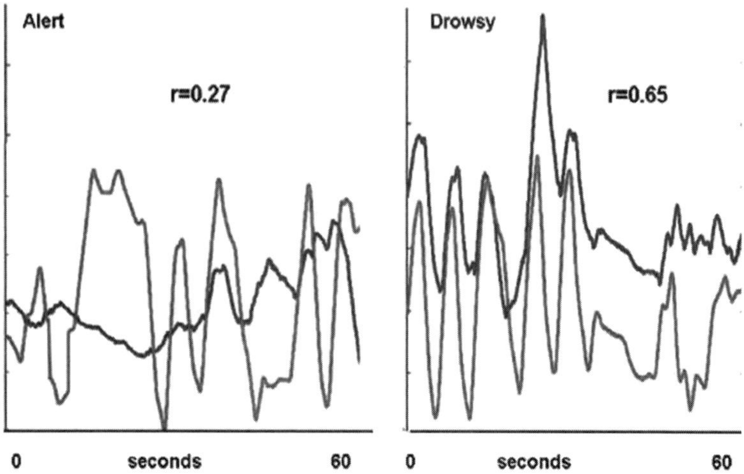

Fig. 8.7 Head motion (*gray*) and steering position (*black*) for 60 s in an alert state (*left*) and 60 s prior to a crash (*right*). Head motion is the output of the roll dimension of the accelerometer

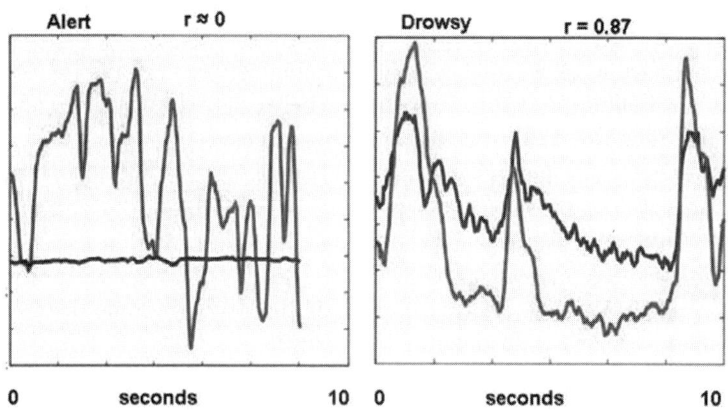

Fig. 8.8 Eye openness and eye brow raises (AU2) for 10 s in an alert state (*left*) and 10 s prior to a crash (*right*)

8.4 Conclusions and Future Work

This chapter presented a system for automatic detection of driver drowsiness from video. Previous approaches focused on assumptions about behaviors that might be predictive of drowsiness. Here, a system for automatically measuring facial expressions was employed to datamine spontaneous behavior during real drowsiness episodes. This is the first work to our knowledge to

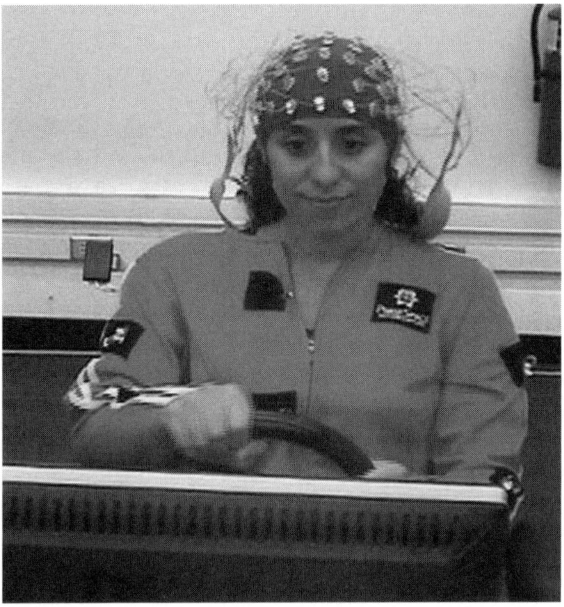

Fig. 8.9 Future experimental setup including the EEG and motion capture capabilities

reveal significant associations between facial expression and fatigue beyond eyeblinks. The project also revealed a potential association between head roll and driver drowsiness and the coupling of head roll with steering motion during drowsiness. Of note is that a behavior that is often assumed to be predictive of drowsiness, yawn, was in fact a negative predictor of the 60-s window prior to a crash. It appears that in the moments before falling asleep, drivers yawn less, not more, often. This highlights the importance of using examples of fatigue and drowsiness conditions in which subjects actually fall sleep.

In future work, we will incorporate motion capture and EEG facilities to our experimental setup. The motion capture system will enable analyzing the upper torso movements. In addition the EEG will provide a ground-truth for drowsiness. The new sample experimental setup can be seen in Fig. 8.9.

Acknowledgments This research was supported in part by NSF grants NSF-CNS 0454233, SBE-0542013 and by the European Commission under Grants FP6-2004-ACC-SSA-2 (SPICE) and MIRG-CT-2006-041919.[4]

[4] Any opinions, findings, and conclusions or recommendations expressed in this material are those of the author(s) and do not necessarily reflect the views of the National Science Foundation.

References

1. DOT: Intelligent vehicle initiative. United States Department of Transportation. www.its.dot.gov/ivi/ivi.htm./
2. DOT: Saving lives through advanced vehicle safety technology. USA Department of Transportation. www.its.dot.gov/ivi/docs/AR2001.pdf
3. Takei, Y., Furukawa, Y.: Estimate of driver's fatigue through steering motion. In Man and Cybernetics, 2005 IEEE International Conference, Volume 2, pp. 1765–1770, 2005.
4. Igarashi, K., Takeda, K., Itakura, F., Abut: Chapter 17 in *DSP for In-Vehicle and Mobile Systems*, H. Abut, J.H.L. Hansen, and K. Takeda (eds), Springer Science-Business, New York, 2005
5. Cobb, W.: *Recommendations for the Practice of Clinical Neurophysiology*, Elsevier, Amsterdam, 1983.
6. Hong, J.E., Chung, K.: Electroencephalographic study of drowsiness in simulated driving with sleep deprivation. *Int. J. of Ind. Ergon.*, 35(4), pp. 307–320, April 2005.
7. Gu, H., Ji, Q.: An automated face reader for fatigue detection. In FGR, pp. 111–116, 2004.
8. Zhang, Z., shu Zhang, J.: Driver fatigue detection based intelligent vehicle control. In ICPR '06: Proceedings of the 18th International Conference on Pattern Recognition, Washington, DC, USA, IEEE Computer Society, pp. 1262–1265, 2006.
9. Gu, H., Zhang, Y., Ji, Q.: Task oriented facial behavior recognition with selective sensing. *Comput. Vis. Image Underst.* 100(3), pp. 385–415, 2005.
10. Bartlett, M., Littlewort, G., Frank, M., Lainscsek, C., Fasel, I., Movellan, J.: Automatic recognition of facial actions in spontaneous expressions. *Journal of Multimedia* 1(6), pp. 22–35.
11. Orden, K.F.V., Jung, T.P., Makeig, S.: Combined eye activity measures accurately estimate changes in sustained visual task performance. *Biol. Psychol.*, 52(3), pp. 221–240, April 2000.
12. Hager, Joseph, C., Ekman, Paul, Friesen, Wallace,V.: *Facial Action Coding System*. Salt Lake City, UT: A Human Face 2002.
13. Kanade, T., Cohn, J., Tian, Y.: Comprehensive database for facial expression analysis. In Proceedings of the Fourth IEEE International Conference on Automatic Face and Gesture Recognition (FG'00), Grenoble, France, pp. 46–53, 2000.

9
Extraction of Pedestrian Regions Using Histogram and Locally Estimated Feature Distribution

Kenji Mase, Koji Imaeda, Nobuyuki Shiraki, and Akihiro Watanabe

Abstract We propose a novel method for extracting target objects such as pedestrians from a given rough initial region of infrared video images from automobiles. The detection of pedestrians from cars is imperative for predicting their subsequent movements in safe driving assistance. However, the automatic extraction of pedestrian regions with various clothing styles and postures is difficult due to the complexities of textural/brightness patterns in clothing and against background scenes. We approach this problem by introducing an object extraction method that reveals the object boundaries while assimilating the variation of brightness distribution. The proposed method can correctly extract the object area by introducing a brightness histogram as a probability density function and an object shape distribution map as a priori probability for the preprocess of the kernel density estimator. We first confirm that the accuracy and computation speed in general cases are improved in comparison to those of the original extractor. We also extend the object distribution map to reflect the shapes of pedestrians with various postures in real situations. Preliminary experimental results are given to show the potential of our proposed method for extracting pedestrians in various clothing, postures, and backgrounds.

Keywords Pedestrian extraction · Posture estimation · Kernel density estimator · Infrared image · Probability density function · Template matching · Bayesian discrimination · Category estimation · Shape-map · Probability map template · Complex background · Complex clothing style

9.1 Introduction

Development of an automatic people locator is one of the highest priorities for automobile driving assistance and autonomous robot navigation. As stated in Chapters 1 and 2, collision avoidance by locating pedestrians is an

---

K. Mase (✉)
Information Technology Center and Graduate School of Information Science, Nagoya University, Nagoya, Japan
e-mail: mase@nagoya-u.ac.jp

K. Takeda et al. (eds.), *In-Vehicle Corpus and Signal Processing for Driver Behavior*, DOI 10.1007/978-0-387-79582-9_9,

important function for driver assistance. Many approaches to pedestrian detection from visible and infrared images [1–3] have demonstrated good results for restricted imaging conditions, e.g., simple background and straight-standing posture. The problem becomes complicated when the illumination changes, the contrast of the object to the background is low, or the observed individuals are making various postures such as crouching. This is partly because the current approaches rely on the edge features of the object region or strong assumptions on the object's contrast. We introduce a novel method that does not rely on the above information; instead it is based on the brightness distribution and shape probability density function. By employing the kernel density function estimation of the brightness distributions of object and background, we can exploit the general probability density features of the object/background regions as well.

Object extraction is one of the most important and difficult issues in image processing. There are many extraction methods, including semi-automatic ones in which a rough region of an object is manually given first to automatically extract a precise object area. Such a semi-automatic method is useful for extracting a target from a photo by manually drawing an initial region, for example. We can use a similar technique with a conventional rough region extraction method such as stereo vision as preprocessing and then apply such an essentially semi-automatic object extraction method to obtain the precise shape of the target. We propose a technique to improve the kernel density estimator, which is one semi-automatic object extraction method, to obtain more accurate results.

We introduce a Bayesian discriminant function as preprocessing for the kernel density estimator to classify an initial region into object and background regions based on the brightness histogram. Thus, we can supply general brightness-based information, i.e., the contrast between object and background regions, to the density estimator. We confirmed the performance of our proposed method through experimental tests with visible and infrared images in actual settings, showing that the accuracy of object extraction and computational speed are improved over the results of the original kernel density estimator. Moreover, our method has another advantage: controlling a priori probability map of the extracted object based on the expected shape of the target object. The map can be generated as a template from the learning samples to form, for example, an arbitrary posture for extracting pedestrians in various postures. The performance of the template map is also demonstrated by experimental results.

In the rest of the chapter, we first survey the related researches. Then we explain the proposed method, followed by preliminary experimental results with general shape objects and the template shape-map adaptation to pedestrian images in real situations. We conclude with discussion and future directions.

9.2 Related Research

There has been much research on pedestrian extraction for automobile applications. Zhao et al. [4] discriminated pedestrians from other objects by applying neural network pattern recognition to the extracted region by stereo depth queue. Fang et al. [5] used a far-infrared image to separately extract the upper and lower bodies. Bertozzi et al. [6] used a multi-resolution image with symmetry and edge density features. Soga et al. [3] used infrared edge features obtained by a Prewitt operator against a stereo range image to achieve good detection performance for straight-standing postures. It is convenient to separate the front object from the cluttered background image with stereo vision [7]. Since this can only provide a rough region of the object, we still need a refining process to obtain an accurate region of an object to detect it as a pedestrian.

The methods of extracting objects from images can be categorized into two approaches. One adjusts a given contour and/or a region by active contour models [8, 9]. The other estimates and determines the category of each pixel, such as Interactive graph cuts [10] and grab cuts [11]. The former represents an active contour as a parametric curve, such as a Spline curve and piecewise line segments. By defining some energy functions as a constraint based on edge strength, smoothness of curves, etc., the given contour is recursively adjusted to fit the curve to a reasonable contour. However, the resulting contour does not closely fit the small, high-curvature, or concave curves due to the smoothness constraint.

On the other hand, the latter pixel-based extraction method uses the probability density functions (p.d.f.) of an object and background regions of each to represent the image's brightness distribution. p.d.f. is often estimated from a brightness histogram [10] or a mixture Gaussian model [11]. By labeling each pixel with the category that gives a higher p.d.f. value, the object region is extracted. This approach outperforms the active contour model in terms of extraction performance for low and high-curvature objects. However, since it uses identical p.d.f.s independently of the pixel's location in the image, category estimation sometimes fails when the brightness distribution of the pixel neighborhood is closer to the opposite category. It also fails when the initial region is given to the extent of providing accurate p.d.f.s.

Takeshima et al. [12] proposed an object extraction method based on a kernel density estimator [11, 14] to solve these problems. It uses the joint probability function of multiple random variables for the position, color, and category of each pixel. The use of random variables of pixel position effectively reduces the estimation error for similar color regions and has robustness against the ambiguity of initial region specification. However, problems remain when the initial object region contains a background edge. We propose introducing a brightness histogram of the initial region as a p.d.f. with an object distribution map as the a priori probability. This is applied to the preprocessing of object extraction of a kernel density estimator. A general survey for the visual analysis of human movement is given by Gavrila [15].

9.3 Kernel Density Estimator With Bayesian Discriminant Function

We attempted to improve the kernel density estimator by introducing a method based on Bayesian discrimination as preprocessing for the initial estimation of an object's region. We assume in this chapter that the initial region is given as a rectangular bounding box, which is manually given in the experiment in the next section. Such a bounding box can be obtained easily from the stereo vision range image in real applications.

9.3.1 Region of Interest for Processing

The region of interest (ROI) is set based on the size and location of the bounding box. In our experiments, we assume that the outskirts of the bounding box represent the background. The ROI is set by enlarging the width and height by 1.5 times the original bounding box to include enough pixels of the background (see Figs. 9.1 and 9.2).

9.3.2 Modifying Probability Distribution Function

We introduce a p.d.f. as an existing-object probability map to give prior knowledge of object position. This is considered useful for reducing estimation error when there is an area of similar brightness between the object and the background.

In this case, we first introduce a simple two-dimensional Gaussian map with a peak in the middle of the ROI and standard deviation computed from its

Fig. 9.1 Input image and bounding box as a given initial region (*rectangle*)

Fig. 9.2 Bounding box (*rectangle*) and enlarged ROI

width and height (Fig. 9.3). That is, the p.d.f. $P_{\mathrm{obj}}(i,j)$ of an object at a pixel (i,j) in ROI I is given by the following:

$$IP_{obj}(i,j) = \frac{1}{\sqrt{2\pi}\sigma_{\mathrm{width}}}\exp\left(-\frac{(i - I_{\mathrm{width}}/2)^2}{2\sigma_{\mathrm{width}}}\right) \times \frac{1}{\sqrt{2\pi}\sigma_{\mathrm{height}}}\exp\left(-\frac{(i - I_{\mathrm{height}}/2)^2}{2\sigma_{\mathrm{height}}}\right), \tag{9.1}$$

Fig. 9.3 Probability distribution function for ROI

where I_{width} and I_{height} are the width and height of the ROI, respectively, and σ_{width} and σ_{height} are set as $\sigma_{\mathrm{width}} = I_{\mathrm{width}}/4$ and $\sigma_{\mathrm{height}} = I_{\mathrm{height}}/4$ so that the region of 2σ will fit within the ROI.

The background probability is given by P_{bg} $(= 1 - P_{\mathrm{obj}})$ for each pixel. To maintain the total equality of object probability over the ROI, we need to stretch the density function to obtain a normalized existing-object probability map. The stretching factor is determined to obtain value P_{obj} between 0.1 and 0.9.

9.3.3 Category Estimation by Histograms

We first roughly estimate the object and background regions with the bounding box and the ROI. We compute a brightness histogram within the bounding box as an object-region brightness model (object model) and another histogram within the ROI, excluding the bounding box region, as a background-region brightness model (background model). The object model is represented by $p(Y_{ij}|w_{\mathrm{obj}}$ with brightness $Y(i,j) = Y_{\mathrm{ij}}$ as the random variable.

Then, discriminant functions $g_{\mathrm{obj}}(Y_{ij})$ and $g_{\mathrm{bg}}(Y_{ij})$ for object and background regions, respectively, can be represented by the following:

$$
\begin{aligned}
g_{\mathrm{obj}}\left(Y_{ij}\right) &= p\left(Y_{ij}|w_{\mathrm{obj}}\right) \times P_{\mathrm{obj}}(i,j) \\
g_{\mathrm{bg}}\left(Y_{ij}\right) &= p\left(Y_{ij}|w_{\mathrm{bg}}\right) \times P_{\mathrm{bg}}(i,j),
\end{aligned}
\tag{9.2}
$$

where $P_{\mathrm{bg}} = 1 - P_{\mathrm{obj}}$. Afterward, we can compute the values of discriminant functions $g_{\mathrm{obj}}(Y_{ij})$ and $g_{\mathrm{bg}}(Y_{ij})$ to estimate the category of pixel (i, j). This category is given by the function that provides the larger value (Fig. 9.4).

9.3.4 Kernel Density Estimation with Proposed Preprocessing

Category estimation is performed by kernel density estimator [12] with the estimated preprocessing results of the initial category, as illustrated in

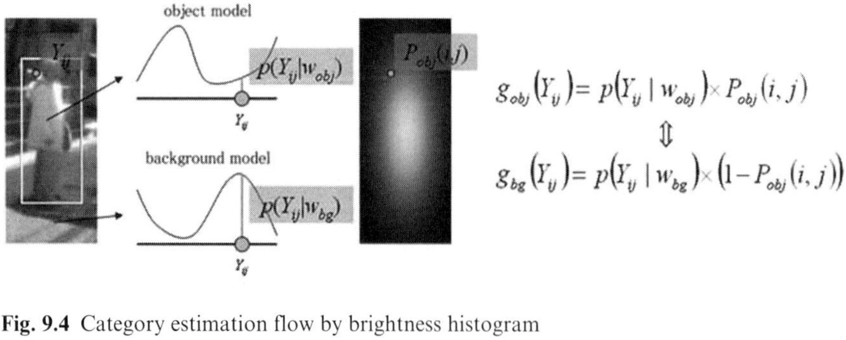

Fig. 9.4 Category estimation flow by brightness histogram

Fig. 9.5 Preprocessing result by histogram matching

Figs. 9.5 and 9.6. The estimated category is used as the initial values of a random variable. We also use the position and the brightness of each pixel and the brightness variance over a neighborhood of 5 by 5 pixels to represent textual features.

The pixel position's bandwidth for kernel density estimation is 1/10 of the ROI's width and height with a uniform kernel, while the brightness bandwidth is 1/2 of the standard deviation of the pixel brightness within the ROI with an Epanechnikov kernel, and the brightness variance bandwidth is 1/2 of the brightness variance across the pixels within the ROI with a Gaussian kernel. These parameters and kernels were chosen heuristically based on a pre-experiment with learning samples. Estimation iteration stops when category estimation is fixed at all pixels.

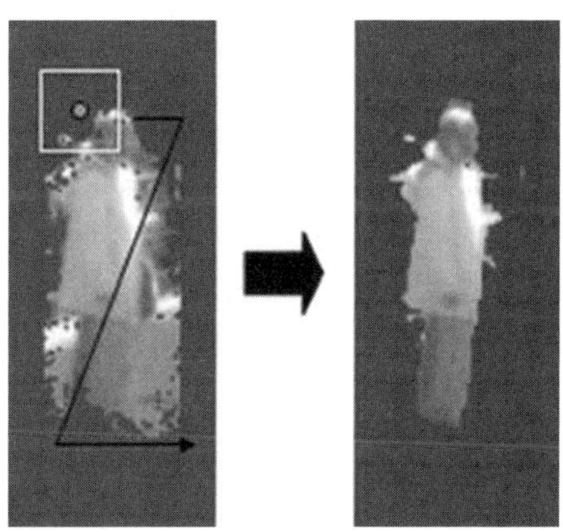

Fig. 9.6 Object region estimated by kernel density estimator with proposed preprocessing

9.4 Pre-Experiment With Gaussian Shape Distribution

We have conducted a general experiment with the original kernel density estimator and the proposed method. Identical test images and bounding boxes are given for each method. The test images are snow (93 × 87 pixels), cup (96 × 84), drink (90 × 90), pedestrian 1 (59 × 138), pedestrian 2 (59 × 136), and pedestrian 3 (58 × 139) (see Fig. 9.7a).

The results are listed in Table 9.1 and shown in Fig. 9.7. The number of error pixels in Table 9.1 represents the sum of true-negative and false-positive pixels. The ground-truth is given by manual drawings over images. The total number of error pixels shows a decreasing trend. For simple objects such as the snow image, the extraction quality of the proposed method is as good as the original.

The proposed method reduces the iteration time, i.e., the computational cost. For such complex shapes as the cup, drink, and pedestrian 1 images, the extraction results with the proposed method are better than the original. However, for such difficult examples as the images of pedestrians 2 and 3, the proposed method fails to extract some parts of the objects because the given object's (a priori) probability map from the p.d.f. assumes a general two-dimensional Gaussian distribution. Consequently, the true object region near the bounding box tends to be classified as background.

9.5 Shape-Dependent Probability Map Template

We now introduce an a priori probability map *(shape-map)* to reflect the target object shape, i.e., a pedestrian in object extraction. The shape-map is generated from the example images of pedestrians from a car-mounted camera. The shape-map in obtained by taking the average of each position's pixel values and then blurred by a Gaussian filter (5 × 5 pixels) after normalizing the width and height of the manually extracted pedestrian shapes.

A preliminary experiment with a pedestrian shape-map was conducted. Figure 9.8 shows some good results obtained after adjusting the parameters of the bandwidths. A better result was achieved by reflecting the target shape in the shape-map for these examples. However, if the shapes of the map and the target object do not match, extraction may fail easily. Since our goal is extracting various postures, we need to use a certain number of shape-maps, each of which represents a class of postures.

At this point in the discussion, we need some criteria to evaluate the matching degree of the map to pick the best match. Using such criteria, we can evaluate the success of shape extraction against a shape-map and determine which is the closest and the most appropriate for the target. We therefore newly define an extraction criterion:

$$E = \mathrm{num\_error\_pixels}(\mathrm{binary}(\mathrm{Shape} - \mathrm{Map}, \mathrm{th}), \mathrm{binary}(\mathrm{ExtractedRegion})), \quad (9.3)$$

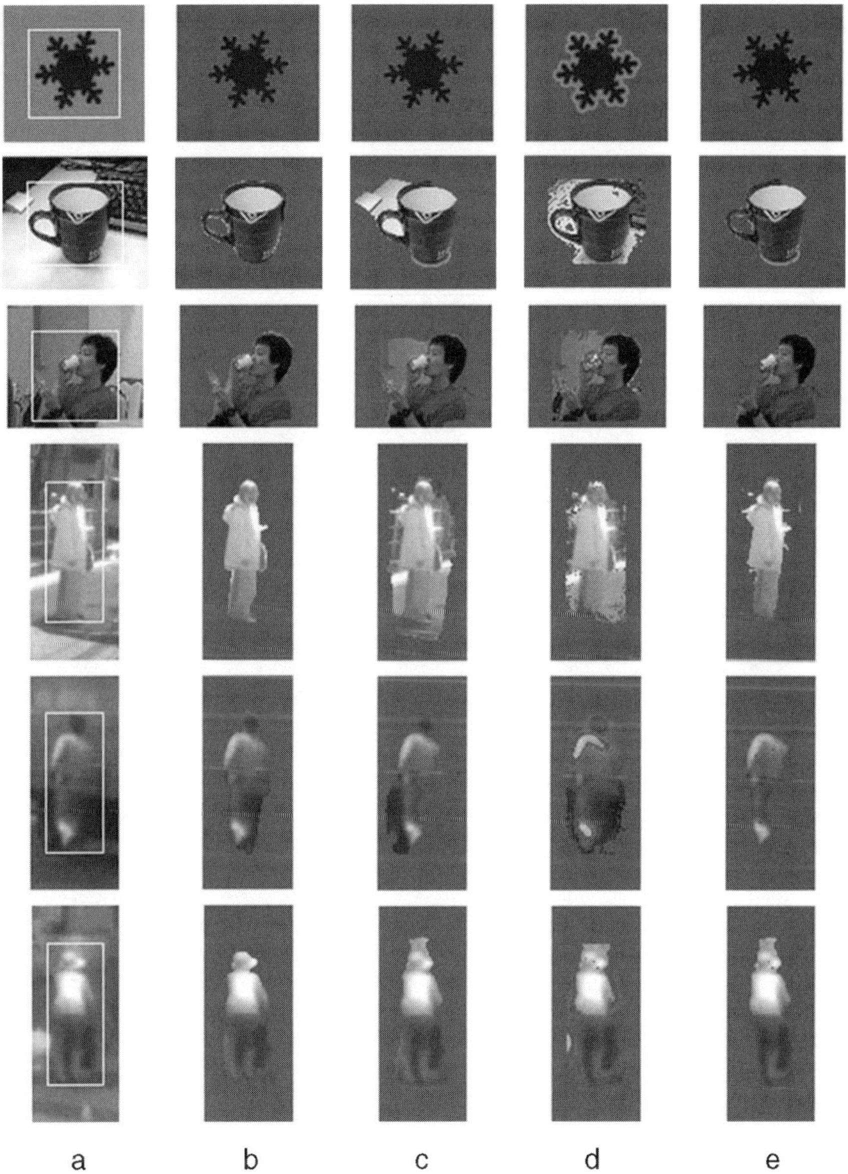

Fig. 9.7 Results of general experiment: **a** Input image and bounding box, **b** ground-truth (manual), **c** original method, **d** preprocessing result by histogram, **e** proposed method. From *top* to *bottom* snow, cup, drink, pedestrians 1, 2, and 3

where *binary(image, threshold)* is a binary function that binarizes an *image* with a threshold, *Shape-Map* is the shape map image, and the *Extracted-Region* is also an image of the extraction result. The *num_error_pixels(A, B)* is a function

Table 9.1 Experimental results: original and proposed methods (*or* original, *pr* proposed)

	Error pixels (%)		Iterations #	
	OR.	PR.	OR.	PR.
Snow	21 (2)	21 (2)	15	8
Cup	727 (37)	233 (12)	19	15
Drink	706 (30)	333 (14)	31	66
Pedestrian 1	1781 (101)	311 (14)	22	63
Pedestrian 2	1137 (60)	821 (43)	85	55
Pedestrian 3	510 (25)	690 (33)	29	16

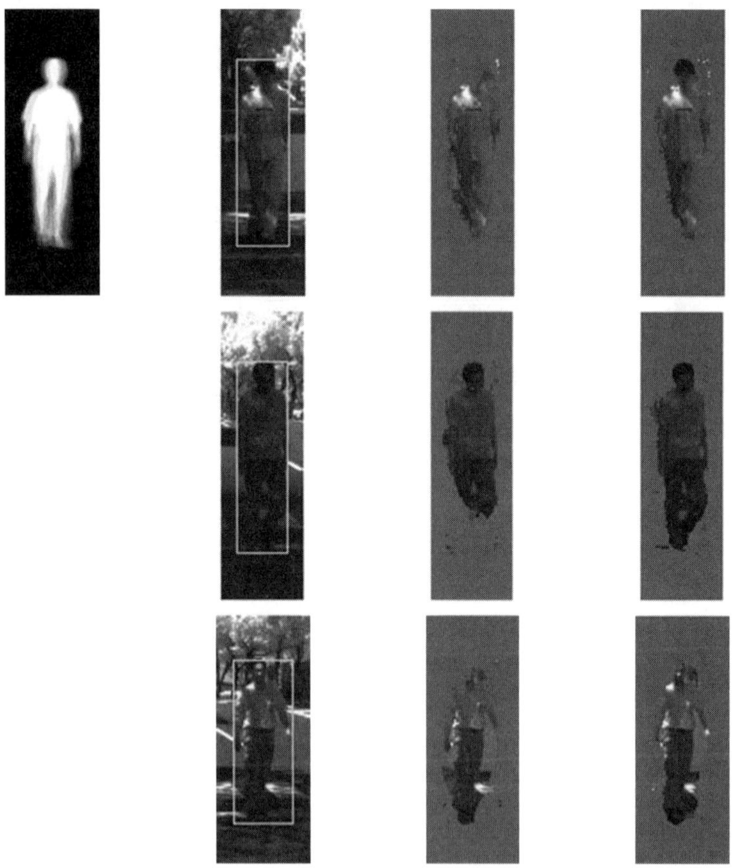

a b c d

Fig. 9.8 Extraction results with a shape-map of frontal pedestrian shape-map: **a** Priori probability map (shape-map), **b** input images and their bounding box, **c** original kernel method, and **d** proposed method

that counts the number of pixels of each value that are "1 (true)" when taking an exclusive OR of two pixel values at the same position of binary images A and B. E is a criterion that reflects our previous discussion of experimental results illustrated for previous experiments in Table 9.1. The value of E equals the number of error pixels, i.e., the sum of true-negative and false-positive pixels. The number is reduced when the extraction performs better.

9.5.1 Experiment of Criterion Performance

We conducted an experiment to determine the performance of matching criteria E by measuring its values as we moved the shape-map along the targeted object position. We set threshold = 0.5 in this experiment and used two kinds of shape-maps: frontal-view (Fig. 9.9a) and side-view (Fig. 9.9b).

The two shape-maps are applied to a test image of a frontally walking pedestrian, as shown in Fig. 9.10. The algorithm shows a better result for the frontal-view shape-map (Fig. 9.10b) than for the side-view shape-map (Fig. 9.10c). These results are obtained by changing the position of the shape-maps along the target

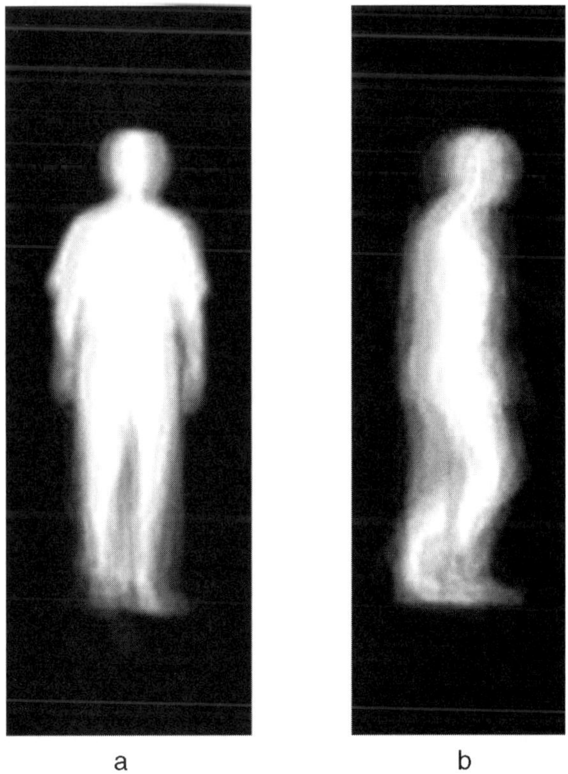

Fig. 9.9 Pedestrian-shaped maps manually composed from original images: **a** frontal-view shape-map and **b** side-view shape-map

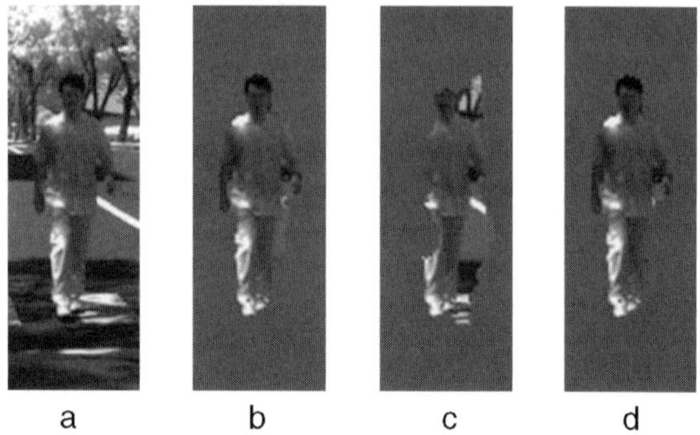

Fig. 9.10 Example of different posture shape-map performances: "similar posture" shape-map (frontal, walking) gives better result. **a** Input test image, **b** minimal *E* result by frontal-view map of Fig. 9a ($E = 338$), **c** minimal *E* result by side-view map of Fig. 9b ($E = 390$), and **d** manually chosen best results (by front-view map, $E = 373$)

region and searching for the minimum value of *E*s. The search is performed here thoroughly in a brute force manner.

The minimum *E* values for both are $E_{\min} = 338$ and $E_{\min} = 390$ for the frontal- and side-views, respectively. As shown in the example, we frequently observe cases where the minimum value is smaller when the appropriate shape-map is used on the target. We need to investigate many cases to evaluate the shape-describing capability of value *E*.

We next observe that value *E* becomes smaller when the map is placed on the appropriate position (Fig. 9.11), although it is influenced to some extent by the

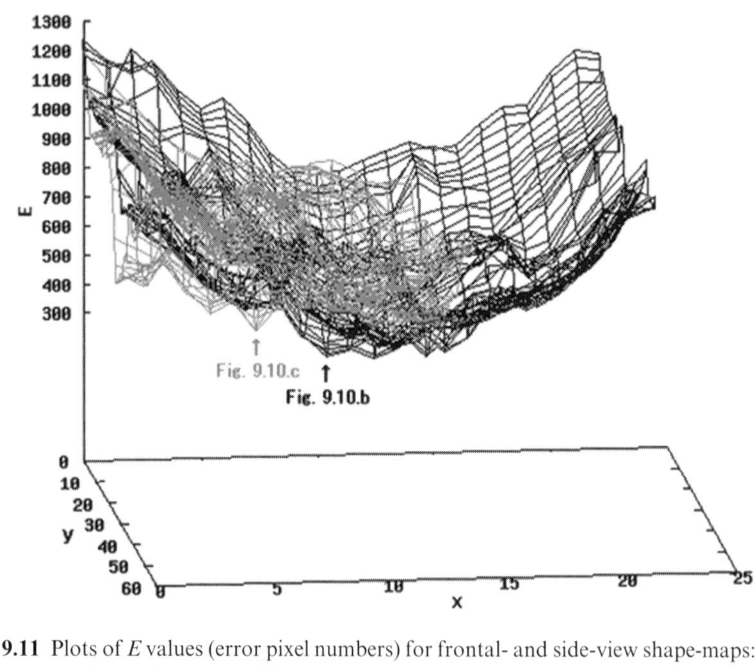

Fig. 9.11 Plots of *E* values (error pixel numbers) for frontal- and side-view shape-maps: *black* and *gray* graphs are front- and side-view maps, respectively

posture and clothing of the target and the edge pattern in the background. The extracting results of the minimum value correspond to Fig. 9.10b and c.

As illustrated here, there are a few valleys (minimum points) of E distribution, meaning that we may fall into a sub-optimal solution during a general fast search with the current definition of E. Actually, even minimum E is obtained for the non-optimum position. Figure 9.10d shows such a case. It has a larger E value, but the extracted shape looks better than Fig. 9.10b. This implies that we have more room to improve the performance by re-defining E. By investigating the details when such cases happen, even a weak edge influences region extraction. E value-based object extraction can be improved to become robust enough against background edges by taking edge information into account.

9.6 Conclusion

We proposed a method that introduces the preprocessing of category estimation to a kernel density estimator to achieve better performance in extracting an object from a given bounding box region. The discriminant function derived from a histogram-based probability density function can give better initial categorization for the kernel estimator. Using the proposed method, computational cost can be reduced for simple objects, and accuracy can be improved for objects with complicated shapes. Since an object's probability map is adjusted in advance by an example object, object extraction becomes sufficiently robust and accurate for pedestrians in various postures. This implies that different shape-maps can be used to estimate the various postures of pedestrians. We showed the capability and performance of our proposed method in a few experiments. Even though our method needs further refinement and improvement, it offers potential to extract regions of various pedestrian postures of different clothing patterns from complex backgrounds. Future issues include a map-generation method, criteria of template selections, and feature selection for more robustness against high-contrast backgrounds.

References

1. D. M. Gavrila, S. Munder, "Multi-cue pedestrian detection and tracking from a moving vehicle," International Journal of Computer Vision, vol. 73, no. 1, pp. 41–59 (2007).
2. I. Haritaoglu, D. Harwood, L. S. Davis, "W4: real-time surveillance of people and their activities," IEEE Transactions on PAMI, vol. 22, no. 8, pp. 809–830 (2000).
3. M. Soga, T. Kato, M. Ohta, Y. Ninomiya, "Pedestrian detection using stereo vision and tracking," IEEE ICDE Workshop on Realworld Multimedia Corpora in Mobile Environment, pp. 1–8 (2005).
4. L. Zhao, C. E. Thorpe, "Stereo- and network-based pedestrian detection," IEEE Transactions on ITS, vol. 1, no. 3, pp. 148–154 (2000).

5. Y. Fang, K. Yamada, Y. Ninomiya, B. K. P. Horn, I. Masaki, "A shape-independent method for pedestrian detection with far-infrared images," IEEE Transactions on Vehicular Technology, vol. 53, no. 6, pp. 1679–1697 (2004).
6. M. Bertozzi, A. Broggi, A. Fascioli, T. Graf, M.-M. Meinecke, "Pedestrian detection for driver assistance using Multiresolution Infrared Vision," IEEE Transaction on Vehicular Technology, vol. 53, no. 6, pp. 1666–1678 (2004).
7. M. Bertozzi, E. Binelli, A. Broggi, M. Del Rose, "Stereo vision-based approaches for pedestrian detection," Proceedings of the 2005 IEEE Computer Society Conference on Computer Vision and Pattern Recognition (CVPRW'05) Workshops, p. 16 (2005).
8. M. Kass, A. Witkin, D. Terzopoulos, "Snakes: active contour models," International Journal of Computer Vision, vol. 1, no. 4, pp. 321–331 (1988).
9. N. Paragios, R. Deriche, "Geodesic active contours and level sets for the detection and tracking of moving objects," IEEE Transactions on Pattern Analysis and Machine Intelligence, vol. PAMI-22, no. 3, pp. 266–280 (2000).
10. Y. Y. Boykov, M. Jolly, "Interactive graph cuts for optimal boundary & region segmentation of objects in N-D images," in Proceedings of International Conference on Computer Vision, pp. 105–112 (2001).
11. A. Blake, C. Rother, M. Brown, P. Perez, P. Torr, "Interactive image segmentation using an adaptive GMMRF model," in Proceedings of the 8th European Conference on Computer Vision, pp. 428–441 (2004).
12. H. Takeshima, T. Kaneko, T. Ida, "Extracting object regions using locally estimated color distributions," Meeting on Image Recognition and Understanding (MIRU2006), IS1-4 (in Japanese) (2006).
13. D. Comaniciu, "An algorithm for data-driven bandwidth selection," IEEE Transactions on Pattern Analysis and Machine Intelligence, vol. PAMI-25, No. 2, pp. 281–288 (2003).
14. M. Singh, N. Ahuja, "Regression based bandwidth selection or segmentation using parzen windows," in Proceedings 338 of IEEE International Conference on Computer Vision, pp. 2–9 (2003).
15. D. M. Gavrila, "The visual analysis of human movement: a survey," Computer Vision and Image Understanding, vol. 73, no. 1, pp. 82–98 (1999).

10
EEG Emotion Recognition System

Ma Li, Quek Chai, Teo Kaixiang, Abdul Wahab and Hüseyin Abut

Abstract This chapter proposes an emotion recognition system based on time domain analysis of the bio-signals for emotion features extraction. Three different types of emotions (happy, relax and sad) are classified and results are compared using five different algorithms based on RVM, MLP, DT, SVM and Bayesian techniques. Experimental results show the potential of using the time domain analysis for real-time application.

Keywords Emotion recognition · EEG · Fuzzy neural network · Confusion matrix · RVM · MLP

10.1 Introduction

Emotions accompany us in our daily life, playing a key role in non-verbal communication. Assessing emotions is thus essential to the understanding of human behavior. In order to achieve an intelligent man–machine interface system that recognizes non-verbal information, such as the intentions, emotions and affections of a user, this chapter aims to present an emotion recognition system by means of pattern recognition and classification techniques.

Human electroencephalography (EEG) measures both the frequency and amplitude of electrical activity generated from the human brain. The benefits of using EEG to conduct experiments testing are noninvasive, simple, fast and inexpensive. It is neither painful nor uncomfortable or time-consuming for the subjects. For these reasons, EEG has become a preferred method in studying the brain's responses to emotional stimuli.

A. Wahab (✉)
School of Computer Engineering, Nanyang Technological University, Nanyang Avenue, Singapore, 639798
e-mail: asabdul@ntu.edu.sg

K. Takeda et al. (eds.), *In-Vehicle Corpus and Signal Processing for Driver Behavior*, DOI 10.1007/978-0-387-79582-9_10,

In this chapter, four popular neural networks from WEKA toolbox have been used on the EEG bio-signals' time domain data in order to determine the optimal selection of experiments parameters.

By considering the results from different neural networks, the sampling length, overlapping rate of sampling and the initial length of signals that need to be removed can be obtained. Using these methods, the selection of experiment parameters is usually more robust and informative.

Finally, experiments are conducted to find important features that classify the three different emotions.

The chapter is organized as follows. After the brief introduction in Section 10.1, Section 10.2 will describe the data collection we are using in the experiments. *Relevance vector machine* (RVM) model is briefly discussed in Section 10.3. After that, the experiments and the analysis of results are presented in Section 10.4. We conclude the whole chapter in Section 10.5.

10.2 Emotional Data Collection

10.2.1 Experimental Setup

Figure 10.1 shows the data collection model for gathering bio-potential signals in emotion recognition experiments. First, subjects are asked to put on the headgear shown in Fig. 10.2 to record their relevant brainwave activities and the

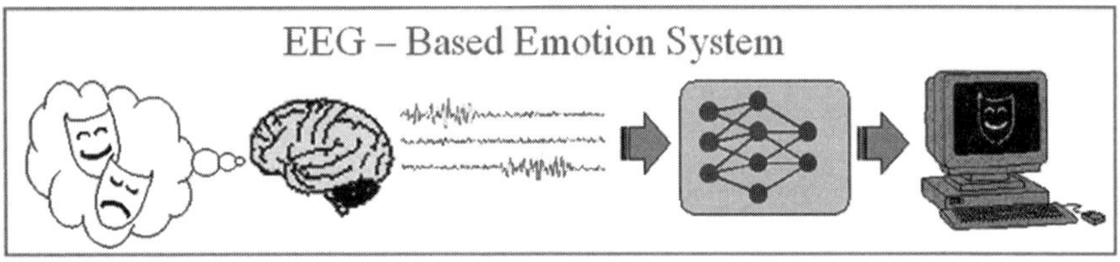

Fig. 10.1 Data collection model for EEG-based emotion detection

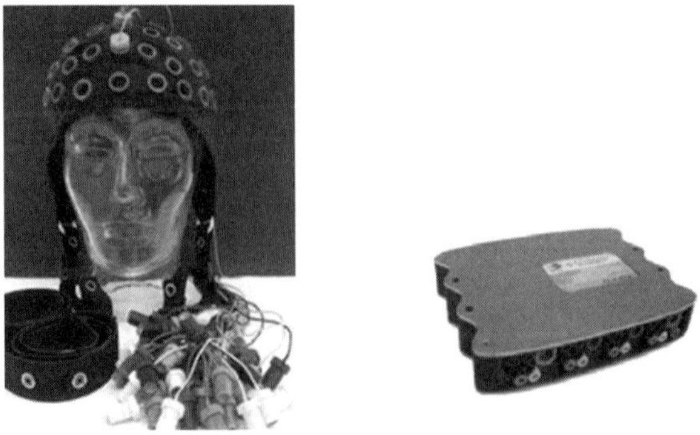

Fig. 10.2 Equipment and subject headgear: MICROelements (*left*) and USBamp (*right*)

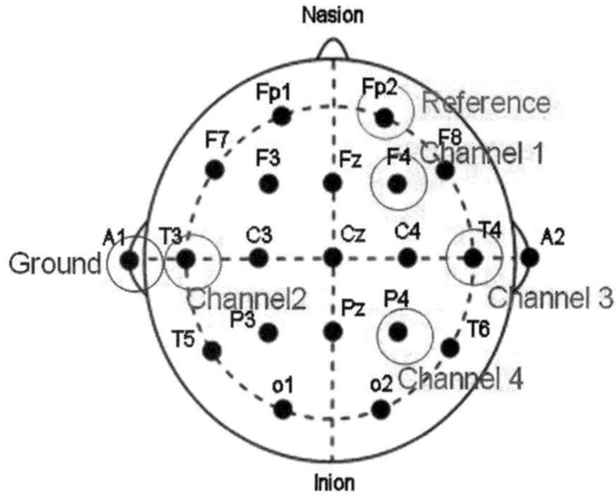

Figure 10.3 Placement of EEG electrodes

EEG device also shown in the same figure. Next these signals are ported to a PC for recording. Another PC is then used to present the visual and aural stimuli to excite the respective emotions of the subject.

Bio-signal data were collected over four dry electrodes shown in Fig. 10.3 from the points F4, T3, T4 and P4 according to the International 10–20 standards. Fp2 was the ground channel and the left ear lobe was used as the reference. The impedances of the electrodes are ensured to be below 40 throughout the experiment.

10.2.2 Psychological Experiments

The experiments were carried out in a laboratory. The EEG bio-signals were gathered under psychological experiments that used visual and aural stimulus for exciting the respective emotions.

Three selected videos from www.youtube.com were used as stimulus for each emotion. A survey was conducted among 30 human subjects who did not participate in the experiments to evaluate the integrity of the videos to invoke the respective emotions among 10 individuals. The results can be seen in Table 10.1. The average results of the three emotions are around 70%. This is

Table 10.1 Emotion survey of human subjects

Subject sample set	18 males	12 females	
Emotion classes	Happy	Relaxed	Sad
Average rating (1–10)	7.07	7.33	7.23

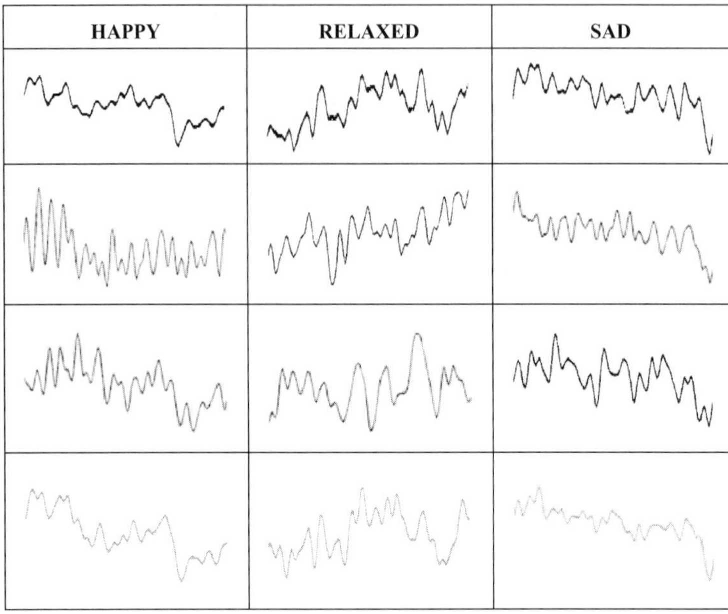

Fig. 10.4 Example of bio-signals: channel 1 (*top row*), channel 2 (*Second row*), channel 3 (*third row*), channel 4 (*fourth row*)

still acceptable as different people have different threshold toward each emotion, which are built over a long time through adaptive learning in uncontrolled environment.

The participants consist of a total of 3 males and 2 females, all native Chinese between the ages of 19 and 25. The raw EEG bio-signals were collected from each subject for each of the three emotions. Each electrode on the head records electrical signals which are then recorded in a channel. Figure 10.4 shows examples of EEG bio-signals measured from a subject while he received the stimulus. Raw EEG data shown in Fig. 10.4 is hard to draw generalization about. Using Matlab, it is possible to build a graphical user interface to the EEG data as well as easily create transformation files to manipulate the EEG data in virtually any way.

10.3 Feature Extraction

In order to make use of neural network, there must be availability of inputs and outputs data for training. This chapter will make use of the feature extraction method proposed in [1] in time domain. According to [1], six features can be extracted in each bio-signal:

$$
\begin{aligned}
\mu_X &= \tfrac{1}{T}\sum_{t=1}^{T} X(t) \\
\sigma_X &= \sqrt{\tfrac{1}{T}\sum_{t=1}^{T} (X(t)-\mu_X)^2} \\
\delta_X &= \tfrac{1}{T-1}\sum_{t=1}^{T-1} |X(t+1)-X(t)| \\
\bar{\delta}_X &= \tfrac{1}{T-1}\sum_{t=1}^{T-1} |\bar{X}(t+1)-\bar{X}(t)| = \frac{\delta_X}{\sigma_X}, \\
\gamma_X &= \tfrac{1}{T-2}\sum_{t=1}^{T-2} |X(t+2)-X(t)| \\
\bar{\gamma}_X &= \tfrac{1}{T-2}\sum_{t=1}^{T-2} |\bar{X}(t+2)-\bar{X}(t)| = \frac{\gamma_X}{\sigma_X}
\end{aligned}
\tag{10.1}
$$

where t, T are the sampling number and the total number of samples, respectively. By using these feature values, a total of 24 features, i.e., 6 for each channel can be determined. No noise-filtering methods are used in preprocessing because our learning model is expected to assist in reducing the noise level.

10.3.1 Selection of Experiment Parameters

Several parameters need to be determined before extracting features from the original dataset, including

1. Sampling length
2. Overlapping rate of sampling
3. Initial length of signal that needs to be removed[1]

For parameter (3), the first 2000 samples in all bio-signals were consistently removed. To determine parameters (1) and (2), a simple experiment was conducted to find out the optimal one.

It can be deduced that for most popular models that were used in the experiment, the trend is that larger the sampling length and the overlapping rate, the better the performance. Thus in the experiment, a sampling length of 5,000 samples and an overlapping rate of 70% are chosen.

10.3.2 RVM Model

As a Bayesian sparse kernel technique, the relevance vector machine (RVM) [2] shares a lot of characteristics with support vector machine (SVM) [3] and

[1] First several seconds of signal recording is invalid as per specifications of the EEG equipment manufacturer.

Gaussian process models (GP) [4]. As discussed in [4] (Section 6.6), the RVM can actually be viewed as a special case of GP, with the covariance function form as

$$k(\mathbf{x},\mathbf{x}') = \sum_{j=1}^{N} \frac{1}{\alpha_j} \phi_j(\mathbf{x})\phi_j(\mathbf{x}'), \tag{10.2}$$

where α_j are hyperparameters and the N basis functions $\varphi_j(x')$ are usually, but not necessarily, assumed to be Gaussian-shaped basis functions centered on each training data point

$$\phi_j(\mathbf{x}) = \exp\left(-\frac{|\mathbf{x}-\mathbf{x}_j|^2}{2l^2}\right), \tag{10.3}$$

where l is a length-scale hyperparameter controlling the width of the basis functions and x_j is the jth input instance.

Due to the similarities of RVM to SVM and GP, there are some advantages of RVM compared to other learning models such as MLP and decision trees. For example, as a special case of GP, RVM can avoid the overfitting by marginalizing rather than cross-validation. In this way, the model selection step can use all the training data, without the need for a validation set. The computation cost in RVM is also much reduced than the models based on cross-validation. For our empirical experience, the RVM model can be several times faster than a MLP based on the 10-folder cross-validation.

Another merit of RVM and SVM is due to their sparse solutions, in which a lot of instances/features play no role [5]. For RVM, by setting different length-scale hyperparameters for different input components, a significant portion of them will go to infinity in the evidence maximization solution, which means the corresponding input component plays no role in the final solution. In this way, RVM can be used as a classifier and also a feature selection method. In the following experiments we will see that only 1/3 features are selected by RVM. And without the non-significant features, the classification performances of all learning models in the experiments did not deteriorate much. However, by only using those significant features, the computation costs have been much reduced.

As the demerits of all kernel methods, we usually need to store the training samples during training of RVM, which may prolong the evaluation time. More details about RVM and GP can be found in [4, 5].

10.4 Experimental Results

In this section, we will be discussing experiments based on four popular neural network models as well as a new learning model based on relevance vector machines (RVM) [1]. The four popular models are multilayer perceptron, decision tree, Bayes network and support vector machine. In all experiments, the accuracies are estimated by the 10-folder cross-validation to avoid

Table 10.2 Confusion matrix for RVM, multilayer perceptron, support vector machine, and Bayes network

	Happy vs relaxed		Relaxed vs sad		Happy vs sad	
RVM	86	9	130	6	95	0
	0	136	1	128	0	129
Accuracy	96.10%		97.36%		100.00%	
MLP	91	4	134	4	93	2
	2	134	7	122	3	126
Accuracy	97.40%		95.85 %		97.77 %	
DT	91	4	126	10	91	4
	5	131	16	113	5	124
Accuracy	96.10%		90.19 %		95.98 %	
SVM	81	14	117	19	80	15
	3	133	22	107	9	120
Accuracy	87.45 %		75.85 %		84.38 %	
Bayes	75	20	108	28	85	10
	1	135	31	98	0	129
Accuracy	90.91 %		77.74 %		95.55 %	

overfitting, which reflects the performance on both training and unseen data and the methods are implemented using WEKA toolbox.

Default settings were all used in WEKA for multilayer perceptron (MLP), decision tree (C4.5) and Bayes network (BayesNet). But for SVM, the buildLogisticModels is set to True, the filterType is set to Standardize training data and the rest uses default values. The resultant confusion matrix of each model in the experiment is listed in Table 10.2. The training time of each model is calculated by taking the average training time of running the experiments five times, which is shown in Table 10.3.

From Table 10.2, it can be seen that both the results of RVM and MLP are relatively close and both give better results than the rest. However, in Table 10.3,

Table 10.3 Training time in seconds for RVM, multilayer perceptron, support vector machine, and Bayes network

	Happy vs relaxed	Relaxed vs sad	Happy vs sad
RVM	60.49 s	74.9 s	145.65 s
Total training time = 281.04 s			
MLP	162.77 s	189.06 s	161.64 s
Total training time = 513.47 s			
DT	0.42 s	0.66 s	0.4 s
Total training time = 1.48 s			
SVM	36.76 s	31.72 s	35.66 s
Total training time = 104.14 s			
Bayes	0.3 s	0.8 s	0.4 s
Total training time = 1.5 s			

Table 10.4 Experimental results of RVM

Happy vs relaxed		Relaxed vs sad		Happy vs sad	
Confusion matrix:		Confusion matrix:		Confusion matrix:	
86	9	130	6	95	0
0	136	1	128	0	129
Total accuracy:	96.10%	Total accuracy:	97.36%	Total accuracy:	100.00%
Features	Importance	Features	Importance	Features	Importance
C11	0.0001	C11	0.0111	C11	0
C12	0	C12	0.0239	C12	0
C13	202.321	C13	71.9024	C13	5.5198
C14	0.0067	C14	0.4269	C14	0.0038
C15	13.2155	C15	0.0351	C15	3.819
C16	0.0018	C16	0.1131	C16	0.0011
C21	0.0001	C21	0.0022	C21	0
C22	0	C22	0.0701	C22	0.0001
C23	0.0656	C23	11.2806	C23	0.016
C24	0.005	C24	0.2564	C24	0.0044
C25	0.0018	C25	0.0829	C25	19.5637
C26	0.0014	C26	0.0939	C26	0.0015
C31	0.0079	C31	0.1388	C31	0.0001
C32	0	C32	0.0002	C32	0.001
C33	5.6478	C33	0.8149	C33	0.0001
C34	0.0037	C34	0.1181	C34	0.0196
C35	9.605	C35	23.2826	C35	0
C36	0.001	C36	0.0621	C36	0.0057
C41	0.0001	C41	0.0021	C41	0
C42	0	C42	0.0003	C42	0.0422
C43	40.2034	C43	74.3512	C43	0.0001
C44	0.0166	C44	0.3375	C44	0.0097
C45	0.0054	C45	1.1293	C45	0
C46	0.0033	C46	0.1283	C46	0.0027

the total training time of MLP is almost 1.5 times more than RVM. This is due to the nature of each model, where RVM can do training and cross-validation at the same time while MLP can only do one at a time. Hence in terms of computational cost, RVM would be the better choice.

Next, we try to identify the useful features out of the 24 features, by using RVM. The experimental results can be shown in Table 10.4.

As by-products, RVM can also give the relative importance of features. Fortunately and amazingly, most of the input features are not significant in our experiment. Besides, the significant features for different classification tasks seem consistent in some degree. In summary, the most important features are

(a) δ_X for ch1
(b) γ_X for ch1
(c) δ_X for ch2 (only for relaxed vs sad)

(d) γ_X for ch2 (only for happy vs sad)
(e) δ_X for ch3 (only for happy vs relaxed)
(f) γ_X for ch3
(g) δ_X for ch4
(h) γ_X for ch4 (only for relaxed vs sad)

We can see that for all channels, only δ_X and γ_X are important, all the other four can be generally ignored—this is quite significant. Among the four channels, ch1 is the most important, after that are ch3 and ch4. Ch2 is partly useful for differentiating relaxed and sad. We will try to verify the feature selection results by comparing the performances with full feature set and those with selected features. The results are shown in Table 10.5.

By comparing the confusion matrix in Tables 10.5 and 10.2, it can be observed that both results are relatively similar. The average accuracy of the 24 original features is 97.82%, while the average accuracy of the 8 selected features is 94.27%. Furthermore, in Table 10.6, the total training time of the 8 selected features is about 4.5 times more than the one with the 24 original features. Hence, these support the idea that the output can be classified by just using the useful features from RVM.

Table 10.5 Confusion matrix based on features selected by RVM

	Happy vs relaxed		Relaxed vs sad		Happy vs sad	
RVM	80	15	133	3	77	18
	0	136	4	125	0	129
Accuracy	93.50 %		97.36 %		91.96 %	
MLP	87	8	131	5	94	1
	3	133	2	127	5	124
Accuracy	95.24 %		97.36 %		97.32 %	
DT	91	4	127	9	91	4
	3	133	13	116	4	125
Accuracy	96.97%		91.70 %		96.43 %	
SVM	74	12	114	22	67	28
	5	131	9	120	13	116
Accuracy	88.75 %		88.30 %		81.70 %	
Bayes	79	19	94	42	87	8
	6	130	29	100	1	128
Accuracy	89.18 %		73.21%		95.98 %	

Table 10.6 Comparison of RVM's training time between selected features and original features

RVM	Happy vs relaxed	Relaxed vs sad	Happy vs sad
24 original features	60.49 s	74.9 s	145.65 s
Total training time = 281.04 s			
8 selected features	12.58 s	16.44 s	32.72 s
Total training time = 61.74 s			

Table 10.7 EEG emotion recognition results in [1]

Emotion class	Accuracy (%)
Joy	58.3
Anger	33.4
Sadness	25.0
Fear	41.7
Relaxed	50.0

Based on [1], an accuracy of 41.7% using SVM is achieved from classifying five types of emotions as summarized in Table 10.7. In this chapter, we use SVM in our experiments too. Based on the 8 selected features, SVM is able to give an average accuracy of 86.25% to classify 3 types of emotions, which are happy, relaxed and sad. The percentage difference in accuracy is almost doubled. This low accuracy in [1] might be because of the additional emotions, anger and fear. At first, the data collection for anger and fear might not be convincing. It is genuinely hard to invoke anger or fear from subjects by showing stimulus, especially anger.

It can be observed that the two emotions with the highest accuracy in Table 10.7 are joy and relaxed, which are considered as positive emotions. On the other hand, anger, sadness and fear are considered to be negative emotions.

According to [6], positive emotions are implemented by more left-hemispheric activity, negative emotions by more right-hemispheric activity. This means there are three negative emotions that generate high activity in the right hemispheric, while two positive emotions generating high activity in the left hemispheric. Perhaps this explains why the accuracy for all the negative emotions is much lower as compared to the positive emotions.

It is known in the literature on this field that happy emotions tend to show strong activities in the left too. Hence, from Table 10.5, it can be observed from RVM and MLP that relaxed vs sad gives the highest accuracy, as it involves classifying a positive and negative emotion.

Sad and happy emotions are known to be associated with distinct subdivisions within the same brain regions. This may explain why an accuracy of 100% in RVM is achieved for happy vs sad by using all the 24 original features, shown in Table 10.2. However, an accuracy of 91.96% is achieved when using the 8 selected features; nevertheless this accuracy is good enough as a trade-off for lesser training time. The speed of the training time is increased by almost 2.5 times.

10.5 Conclusions

This chapter proposes an emotion recognition system from EEG signals in time domain. Three emotions, happy, relaxed and sad, were conducted in this experiment. In order to compare the accuracy results, five different models were used and they are relevance vector machine (RVM), multi layer perception (MLP), decision tree (DT), support vector machine (SVM) and Bayes network. Based on the experiments, RVM and MLP both gave the highest accuracies, with relatively similar results. However, it was concluded that RVM is a better choice over MLP, the reason being the training time of the RVM is 1.5 times faster. RVM also generated the relative importance of features, which identifies eight important features. The experiment was rerun again with these eight selected features to verify the accuracy. The average accuracy is 97.82% for the 24 original features, while the average accuracy for the 8 selected features is 94.27%. It was also observed that the training time for the latter is almost 4.5 times faster than the other one. Considering the trade-off from the accuracy for the increase in speed of the training time, the experimental results were satisfying.

For future work, a comparison between time domain and frequency domain is planned. Additional features for anger and fear emotion classes are to be included, provided the data collection process is more promising. In addition, increasing the number of EEG electrode locations is a possibility to achieve a better understanding.

Acknowledgments This work has been partially supported by NEDO (New Energy and Industrial Technology Development Organization) of Japan. The authors would like to thank Professor Kazuya Takeda of Nagoya University, Japan for his support and timely advises.

References

1. K. Takahashi (2000), Remarks on emotion recognition from bio-potential signals, Proceedings of 2nd International Conference on Autonomous Robots and Agents.
2. M. E. Tipping (2001), Sparse Bayesian learning and the relevance vector machine, *Journal of Machine Learning Research*, vol. 1, pp. 211–244.
3. V. N. Vapnik (1995), *The Nature of Statistical Learning Theory*. New York: Springer.
4. C. E. Rasmussen and C. K. I. Williams (2006), *Gaussian Process for Machine Learning*. Cambridge, MA: MIT Press.
5. C. M. Bishop (2006), *Pattern Recognition and Machine Learning*. Berlin: Springer.
6. T. Isotani, D. Lehmann, R. D. Pascual-Marqui, M. Fukushima, N. Saito, T. Yagyu, T. Kinoshita (2002), Source localisation of brain electric activity during positive, neutral and negative emotional states, International Congress Series 1232.

11
Three-Dimensional Ultrasound Imaging in Air for Parking and Pedestrian Protection

Marco Moebus and Abdelhak Zoubir

Abstract Acoustic imaging has been used in a variety of applications, but its use in air has been limited due to the slow propagation of sound and high attenuation. We address the problem of obstacle detection in a scene using ultrasound imaging with a 2D array under the constraint of a fixed platform. Applications are in the area of autonomous navigation such as a parking car as well as pedestrian detection for pre-crash measures of crash avoidance. The presented system uses a single transmit pulse combined with a beamformer at the receiving array based on a near-field model to obtain 3D images of a scene. Results from experiments conducted in a laboratory demonstrate that it is possible to detect position and edge information from which an object can be reconstructed. Additionally, echo characteristics change with respect to surface texture. Thus, ultrasound arrays can be employed in cars to augment short-range applications.

Keywords Ultrasound imaging · Beamformer · Obstacle detection · Autonomous navigation · Array processing · Near-field model · Texture · Fresnel's approximation · Huygens' principle · Time-of-flight (TOF) · Additive white Gaussian noise (AWGN)

11.1 Motivation – Why Ultrasound?

While there are a manifold of sensor modalities used in modern cars, the use of ultrasound (US) has been limited to simple distance measurements in parking scenarios. However, as demonstrated by several species in nature, US signals can be employed to obtain far more than range information. Using US arrays, it is possible to exploit spatial information regarding the direction-of-arrival (DOA), effectively creating an *acoustic camera* which operates based on

M. Moebus (✉)
Institute of Telecommunications, Technische Universität Darmstadt, Merckstraße 25, 64283 Darmstadt, Germany
e-mail: moebus@ieee.org

K. Takeda et al. (eds.), *In-Vehicle Corpus and Signal Processing for Driver Behavior*, DOI 10.1007/978-0-387-79582-9_11,

sound waves instead of light. US arrays have been successfully used for imaging in medical, material or underwater sonar applications [5–10].

Their use in air is still limited due to physical restrictions such as slow speed of propagation and high attenuation as well as the presence of a number of competing sensor modalities.

However, for *short-range* applications of a mobile platform like a car, they offer a cheap, reliable and low-power alternative to optical or lidar/radar sensors and can provide discrimination between close objects and background (as opposed to optical sensors). This can be highly useful in reverse parking scenarios or pedestrian detection (e.g., for pre-crash measures). More generally, an US array provides an acoustic camera which can be employed generally for object detection and, to some extent, classification, e.g., for scene reconstruction [2, 3].

Due to the specular scattering characteristics of ultrasound, most existing US array imaging systems analyze an object by moving the platform around it. This procedure records a larger number of scattering samples, e.g., for applications such as underwater exploration or ultrasonography. Nevertheless, in many cases it is desirable to analyze objects or a whole scene of objects independent of the movement of the sensors. This holds especially true for applications where the platform trajectory is independent of the sensor system, such as a parking car, as well as applications where the platform is totally static.

Additionally, this becomes even more important if the system has to perform in an environment where it is likely to encounter many man-made objects possessing solid, smooth surfaces (in relation to wavelength), which increases the effect of specular scattering. In such scenarios, the scattering has a major impact on the way the obtained images should be interpreted on a higher level.

In this chapter, we present an acoustic imaging system that works in air using a 2D array of ultrasound sensors mounted on a fixed platform. We show that the images created with this system carry information about position and edges of the objects which can be exploited for higher level modeling.

11.2 Signal Model

Before we present the imaging system, we make several assumptions about the scene as well as the signals involved:

- The scene is illuminated by a *narrow-band* ultrasound signal with center-frequency f_C and wavelength λ, emitted from a single ultrasound sensor at a fixed position.
- Echoes are recorded by an N-element dense array of isotropic ultrasound sensors with uniform noise of variance σ_n^2 at each element.
- The array operates in air, i.e., signals propagate in a homogeneous linear medium with constant propagation speed (as opposed to human tissue or water).

- Objects are in the near-field, such that the propagation of the sound echoes can be modeled using Fresnel's approximation.
- Additionally, the objects are assumed to have a solid surface, resulting in large acoustic impedance between air and the objects' materials. As a consequence, nearly no power is penetrating into the object and only hard echoes are present.

These assumptions are motivated by physical or technological constraints in order to obtain a system that is suitable to produce several images per second.

If a signal $s(t)$ is projected from a position $\mathbf{p}_T$ into the scene, the echo impinging on a single receiver element i from a point source at position

$$r(\theta, \phi) = r \begin{pmatrix} sin(\theta)cos(\theta) \\ sin(\theta)sin(\phi) \\ cos(\theta) \end{pmatrix} \tag{11.1}$$

can be modeled in the near-field according to the Fresnel approximation as [1,5]

$$x_i(t) = \frac{1}{||\mathbf{p}_T - \mathbf{r}||^2} C a_i(\mathbf{r}) s(t) e^{j \frac{2\pi}{\lambda} \tau} + n(t), \tag{11.2}$$

where

$$a_i(\mathbf{r}) = \frac{1}{r^2} e^{-j \frac{2\pi}{\lambda} \left(2r - ||\mathbf{r}\mathbf{p}_i|| - \frac{||\mathbf{p}_i||^2}{2r}\right)}, \tag{11.3}$$

C is a factor representing the echo-shaping effects of the object's surface and material and r is the distance from the array center to the object. Here $a_i(\theta, \phi, r)$ represents the phase shift and attenuation that occurs due to the position $\mathbf{P_i}$ of the sensor as well as the position $\mathbf{r}$ of the source, with respect to the coordinate origin (see Fig. 11.1). In addition, τ stands for the time-of-flight (TOF) of the signal between transmission and reception of $s(t)$. The noise $n(t)$ is assumed to be

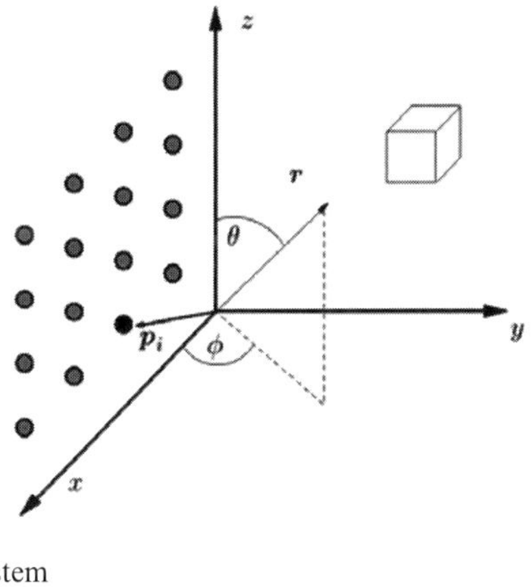

Fig. 11.1 Coordinate system

additive white Gaussian noise (AWGN) with power σ_n^2. As the object's shape is generally unknown and not restricted, the returns from an object can only be modeled rather generically as a superposition of point sources according to Huygens' Principle, assuming a linear channel. In air, this assumption has shown to be a good approximation for narrow-band frequencies below 100 kHz.

11.3 Image Generation

To generate 3D images of a scene in air, one of the limitations is the slow speed of propagation. In contrast to other typical imaging applications, we therefore do not perform beamforming to *transmit* the signal. A better strategy is to illuminate the scene to be analyzed by a single ultrasound source standing at a fixed position near the array. By this, we are able to image the scene by processing the back-scattered reflections from only one transmit pulse and maintain a scan period suitable for a real-time application (see Fig. 11.2).

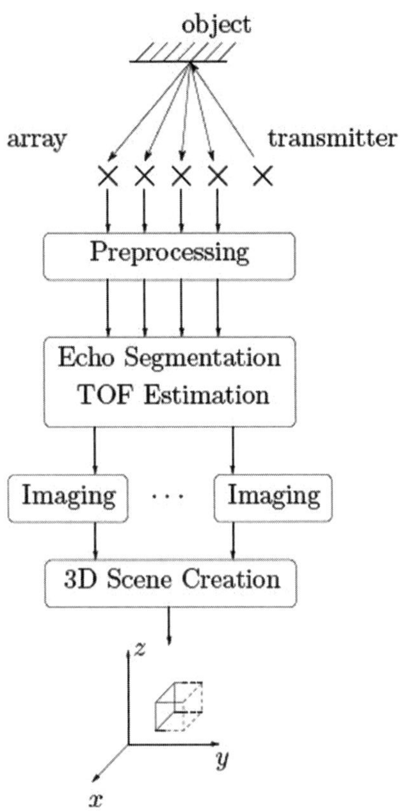

Fig. 11.2 Imaging system flowchart

11.3.1 System Setup

We chose to use a synthetic aperture approach and synthesize the array by a single receiver mounted on a high-precision 2D positioning system in the xz-plane. This implies that the environmental parameters such as object's position and temperature have to be static during the measurements to guarantee reproducibility of the experiments. This can safely be assumed to be true since the synthesis of an array does not exceed a time interval of a few minutes. The synthetic aperture approach also allows analyzing different array layouts for optimization (see [4, 10]). Both the fixed transmitter and the receiver are piezo-electric devices with a membrane of diameter 6.9 mm and a resonance frequency $f_C = 48$ kHz. These are standard ultrasound devices found in bumpers of many cars in the market. The transmitter's membrane is excited by a sinusoidal signal at frequency f_C with a duration of 100 μs. Due to the inertia of the sensor membrane, this results in a narrow-band excitation signal of that frequency and duration of 1 ms. Due to vibration decay, this results in an excitation signal with duration of approximately 1 ms. The received analog signals at the array channels are band-limited before they are sampled at a rate of $f_S = 200$ kHz.

11.3.2 Data Processing

After recording the reflections in each array element, the data is amplified and transformed to an analytic signal in base-band. In order to avoid scanning over a 3D space, one can obtain a range estimate $\hat{r}$ from the TOF for each echo and process them separately in the (θ, ϕ)-space.

To achieve this, we apply matched filtering and obtain a noise estimate by analyzing the signal of one reference sensor up to a time $\tau_{\min}$, which corresponds to the minimal distance $r_{\min}$ of an object. Since no echoes are assumed to be present in this interval, an estimate $\hat{\sigma}_n^2$ of the noise floor is calculated.

It is worth noting that since the noise is assumed to be AWGN, we do not need a large number of samples. It is sufficient to set $r_{\min}$ to a small value (e.g., $r_{\min} = 20$ cm). A threshold can then be based on $\hat{\sigma}_n^2$ in order to identify the start of an echo segment, where echoes have to occur with a minimal duration of 1.0 ms.

These segments are then processed individually by the beamforming algorithm, where $\hat{r}$ is estimated with reference to the start of the echo segment, resulting in a dynamic focusing system. Note that the translation of the TOF into range assumes a direct-path echo. Additionally, note that due to the possible overlap of different reflections, the length of the echo segments might vary. In that case, the later echo is assumed to be in the same range as the first one with the possibility of introducing a small range error for some parts of the analyzed scene. Alternatively, one could process each segment block-wise, which is computationally much more expensive.

11.3.3 Beamforming

In array systems, signal processing crucially determines the overall system performance by affecting the structure of the created images and thereafter object modeling, detection and classification.

In contrast to applications in radar and communications, the imaging application considered in this chapter does not allow to model signal sources as point sources. Unfortunately, this results in the fact that many of the existing adaptive approaches in array signal processing cannot be applied, as they have been developed for far-field conditions and a finite number of *point* sources. In particular, albeit the existence of high-resolution subspace algorithms, these methods cannot be applied, as they assume arrival of signals from distinct DOAs, such that each source is represented in the covariance matrix as a single eigenvalue. In contrast to that, the US imaging system must operate on objects that are close and have a non-negligible spatial spread.

Although some work has been done to extend these methods to spatially spread sources for DOA estimation, the focus is merely on an increase of estimation accuracy [8]. It is not possible to apply these extensions to source imaging. Hence, acoustic imaging systems without prior knowledge about the objects to be imaged have to use beamforming algorithms to create the spatial spectrum. In the following, we compare the effect of two beamforming algorithms in the context of acoustic imaging systems that employ ultrasound in air. We investigate the effects on the created images of the scene and compare these images in terms of artifacts and object visibility.

11.3.4 Evaluation Criteria

The images we obtain by the system described above form the basis for object detection and classification. It is therefore crucial to create images which represent the physical scene in a way that allows for clear discrimination of objects from artifacts. More specifically, the beam pattern should not allow for artifacts due to side lobes while the reflections of the objects should be represented distinctively. As the amount of power present in the ultrasound echoes varies greatly with range, shape and surface of the objects, any variations of sensitivity in the beam patterns over the whole search space are undesirable.

While these criteria are conflicting, the question remains how the beamforming algorithms fulfill those and what differences are present in the resulting images. To base the analysis not only on subjective assessment, we next present objective figures of merit that allow us to compare single image regions numerically.

We compare the beamforming algorithms by the Structured Similarity Index (SSIM) [9]. It is a similarity measure which was introduced in the image processing community as a means of comparing images based on structural information. Additionally to the single quantity of the index, a map showing local similarities between two images can be calculated. Similarity is measured by statistics based on luminance and contrast. As a third component, the structural information of the images is compared by a measure similar to the correlation coefficient, based on pivoted statistics for luminance and contrast.

The index provides both global and local information about the similarity of two images. In our application, we compare the obtained, normalized images to a perfect binary reference image. In addition to the SSIM, we also compute the total power received from the object and non-object regions as well as the variance of the power distribution in the object region to compare how power is received from a single surface. The average power per pixel present in the non-object region is measured to determine the overall strength of the artifacts in the images.

11.4 Experiments and Discussion of Results

The experiments in this section have been conducted in an acoustic laboratory. To exclude effects of the array geometry on the image, all objects have been recorded with a 20×20-element array, resulting in an aperture size of approximately 7×7 cm^2. Similar results are obtained using optimized array geometries with less than 50 sensor elements [4]. All images were obtained using a grid resolution of 1° in both dimensions and are displayed in logarithmic scaling.

11.4.1 Rough Surface Structure–Continuous Response

In this experiment, a circular pole with a diameter $d = 0.185$ m was placed in front of the array. The surface of the object shows a rough structure in the dimension of λ, such that the scattering on its surface is highly affected. As can be seen in Figs. 11.3 and 11.4, the created images show a continuous response over the whole surface, independent of the used beamformer. However, the exact distribution of power as it is received by the array is determined by the beam pattern and therefore the used beamforming algorithm. Artifacts in the images mainly depend on the side lobe structure of the beam pattern. In Fig. 11.3, Capon's beamformer received a fairly constant power level from the whole surface of the pole which fronts the array. Due to its adaptive character, side lobe effects are minimized such that the object is clearly visible whereas the rest of the images only contain artifacts that are more than 10 dB weaker.

For the same object, Bartlett's beamformer shows peaks on the same region, but the response is not as constant as for Capon's beamformer [1]. There are clearly two peaks visible at both ends of the pole along the θ-dimension. The above results can be easily explained by the peak side lobe of the non-adaptive beam pattern of the algorithm. The side lobe effect is also more clearly visible

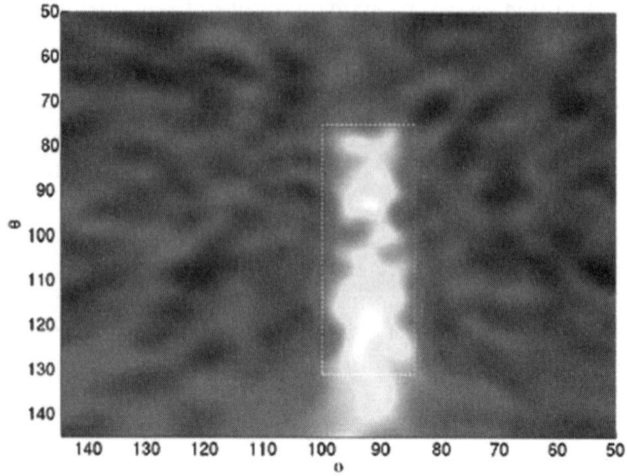

Fig. 11.3 Image of a rough-surfaced object using Capon's beamformer

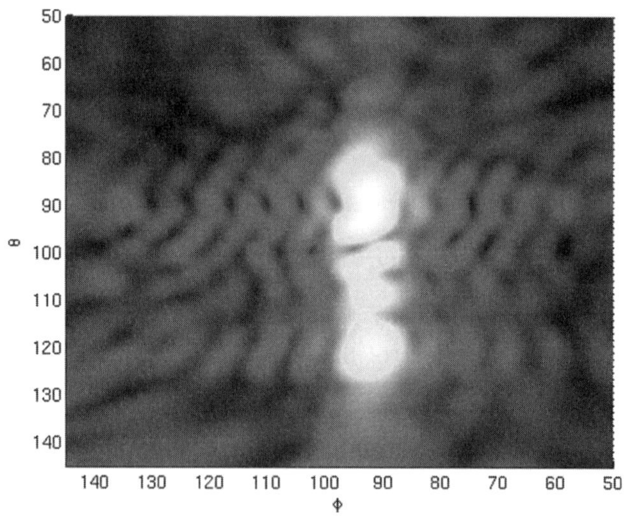

Fig. 11.4 Image of a rough-surfaced object obtained using Bartlett's beamformer

than for Capon's beamformer in non-object regions. Artifacts are stronger because a larger amount of power reflected from the object is captured by side lobes when looking in regions where no object is actually present.

In Table 11.1, the effects in the images are summarized. While the SSIM index is on the same level in the object region (SSIM_O) for both algorithms, Bartlett's beamformer performs worse in the non-object region (SSIM_{NO}). Average power per pixel in the object region (σ_{O}^2) is slightly less for Capon's

Table 11.1 Performance for continuous response

	$SSIM_O$	$SSIM_{NO}$	σ_O^2	$\tilde{\sigma}_O^2$	σ_{NO}^2
Bartlett	0.8886	0.0268	0.1987	$1.3e^{-7}$	0.0460
Capon	0.8852	0.0331	0.1773	$9.3e^{-16}$	0.1247

beamformer than for Bartlett's, whereas the variance of power is significantly less in that region ($\tilde{\sigma}_O^2$). However, comparing overall power in the non-object region (σ_{NO}^2), the Bartlett beamformer performs much better.

11.4.2 Smooth Surface – Specular Response

We have recorded data from another object with the same shape, but a smooth surface. This leads to a highly specular scattering at the object and therefore much lower peak amplitude in the echoes. Additionally, as the reflection on the surface is specular, the array receives echoes only from some characteristic spots on the object such as the direct reflection edges. In such scenarios, the discrimination between artifacts and object echoes is much more difficult. As can be seen in Figs. 11.5 and 11.6, the object is not visible by its whole surface anymore.

Only a direct reflection, ground reflection and a reflection from the upper edge are visible. However, the sources of reflection are much more separated in the image based on Capon's beamformer, as the side lobe effects are stronger than in the previous experiment. Considering the much smaller spatial spread from which power is received, we see that Bartlett's beamformer is incapable of sharply discriminating objects from background here. As shown in Table 11.2,

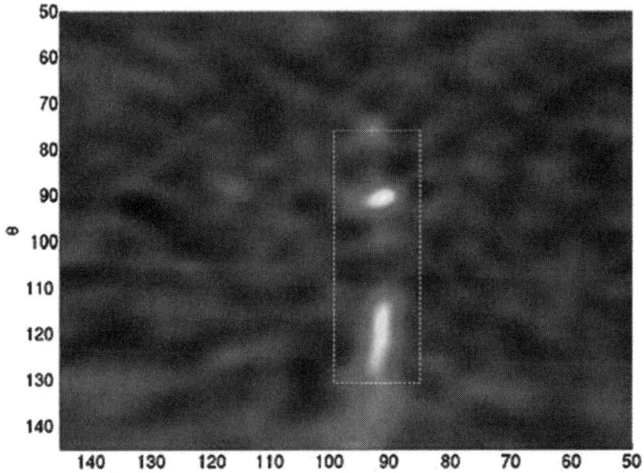

Fig. 11.5 Image of a smooth-surfaced object using Capon's beamformer

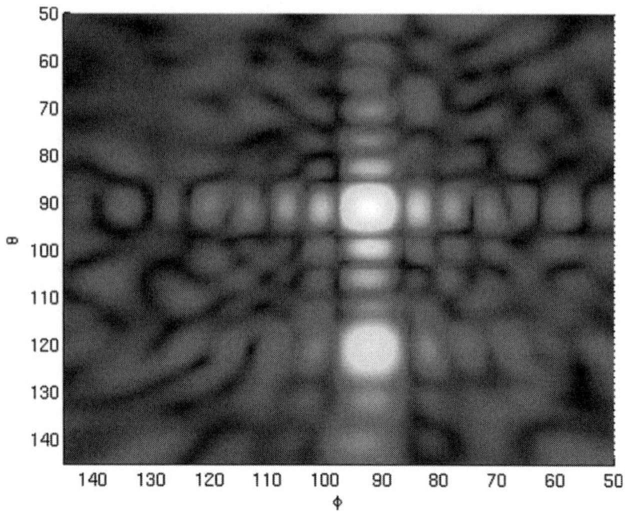

Fig. 11.6 Image of a smooth-surfaced object using Bartlett's beamformer

Table 11.2 Performance for specular response

	$SSIM_O$	$SSIM_{NO}$	σ_O^2	$\tilde{\sigma}_O^2$	σ_{NO}^2
Bartlett	0.8768	0.0371	0.0040	$1.7e^{-6}$	0.0015
Capon	0.8740	0.0425	0.0193	$4.2e^{-12}$	0.00079

the SSIM is similar to the results from the previous experiment. However, due to the specular scattering, sharper peaks occur in the images and the average power per pixel is significantly smaller.

The main difference in this example is that Capon's beamformer outperforms Bartlett's when considering power in the non-object region. Additionally, one can see that Capon's beamformer clearly shows the upper edge of the object and also shows narrower peaks in general. The ground reflection aligned along the symmetry axis is also clearly identifiable.

11.5 Conclusions

In this chapter, we presented a 3D ultrasound imaging system that operates in air and in a fixed position using beamforming. Although the scattering characteristics for solid objects depend on the surface texture, the presented system can be used to obtain images that contain information about edges and the objects' positions on ground. The system is capable to detect objects in the short-range environment of a mobile platform such as a car using a very low cost sensor technology. It is therefore an alternative or enrichment to currently used systems in that context.

The analysis shows that the used beamformer is crucial to the nature of the obtained images and that the higher computational complexity of an adaptive algorithm such as Capon's beamformer is justified by improved artifact behavior as well as spatial resolution consistently for different types of surface textures.

References

1. Krim H, Viberg M (1996) "Two decades of array signal processing", IEEE Signal Processing Magazine, vol. 13, no. 4, pp. 67–94.
2. Moebus M, Zoubir AM (2007) "Three-dimensional ultrasound imaging in air using a 2D array on a fixed platform", Proceedings of the IEEE International Conference on Acoustics, Speech, and Signal Processing (ICASSP '07), Honolulu, USA.
3. Moebus M, Zoubir AM (2007) "A comparison of beamforming algorithms for 3D ultrasound imaging in air", Proceedings of the Biennial on DSP for In-Vehicle and Mobile Systems, Istanbul, Turkey.
4. Moebus M, Zoubir AM (2008) "On the design of sparse arrays using difference sets", Proceedings of the Fifth IEEE Sensor Array and Multichannel Signal Processing Workshop (SAM 2008).
5. Murino V, Trucco A (2000) "Three-dimensional image generation and processing in underwater acoustic vision", Proceedings of the IEEE, vol. 88, no. 12, pp. 1903–1948.
6. Powell D, Hayward G (1996) "Flexible ultrasonic transducer arrays for nondestructive evaluation applications. I. The theoretical modeling approach", IEEE Transactions on Ultrasonics, Ferroelectrics, and Frequency Control, vol. 43, pp. 385–392.
7. Shung K, Zippuro M (1996) "Ultrasonic transducers and arrays", IEEE Engineering in Medicine and Biology Magazine, vol. 15, pp. 20–30.
8. Tapio M (2002) "Direction and spread estimation of spatially distributed signals via the power azimuth spectrum", Proceedings of the IEEE International Conference on Acoustics, Speech, and Signal Processing (ICASSP '02), Orlando, USA.
9. Wang Z, Bovik AC, Sheikh HR, Simoncelli EP (2004) "Image quality assessment: from error visibility to structural similarity", IEEE Transactions on Image Processing, vol. 13, no. 4, pp. 600–612.
10. Yen JT, Steinberg JP, Smith SW (2000) "Sparse 2D array design for real time rectilinear volumetric imaging", IEEE Transactions on Ultrasonics, Ferroelectrics, and Frequency Control, vol. 47, pp. 93–110.

12
A New Method for Evaluating Mental Work Load In *n*-Back Tasks

Naoki Shibata and Goro Obinata

Abstract It is important to evaluate mental work load on drivers will be useful for managing operator's work load or for preventing from overloaded tasks. However, there does not exist any quantitative method for evaluating mental work load in real time. This motivates the study of proposing a new method to evaluate the influence of mental work load caused by information processing demand. Our method focuses on involuntary eye movement of human, which is vestibulo-ocular reflex (VOR). The eye movement occurs reflexively for gaze stabilization while paying attention to the target. We have investigated the influence of mental work load on VOR using a new model-based method. The first step of the method is to identify the eye movement model for a particular subject from measured data without any secondary task. This model represents the subject's dynamics of VOR. After that the eye movement is measured when the subject get distracted by paying attention to secondary tasks, and it is compared with the identified model output. This method makes it possible to assay the influence of mental work load on VOR. This study has investigated the influence of mental work load on human eye movement by giving *n*-back tasks as the secondary task. By varying the amount of information processing demand of *n*-back tasks, we compare the variations of the dynamics of VOR from the identified model which represents human VOR dynamics in ideal situation. Consequently, we give a proof of quantitatively evaluating mental work load using our proposed model-based method.

Keywords Vestibulo-ocular reflex · Mental work load · Model-based approach · Attention · Quantitative evaluation · Verbal *n*-back task · Semicircular canals · Statoliths · Eye movement · Working memory · Coherence · Reaction time · Distraction

---

G. Obinata (✉)
EcoTopia Science Institute, Nagoya University, 464-8603, Nagoya, Japan

K. Takeda et al. (eds.), *In-Vehicle Corpus and Signal Processing for Driver Behavior*, DOI 10.1007/978-0-387-79582-9_12,

12.1 Introduction

Mental workload is an essential concept in most Human-Machine systems. Mental workload can be defined as the currently used amount of cognitive resources in a person at a time point. Since cognitive resources in humans are limited, human performance is easily deteriorated by heavy mental workload. Many methods have been developed to quantify mental workload. These techniques can be categorized in three groups [1]: (a) primary/secondary task measures, (b) subjective-rating techniques including NASA-TLX, and (c) physiological measures. Unfortunately, most existing mental workload measures share some of the major disadvantages, especially when measuring vehicle driver's mental workload.

First, the primary/secondary task measures cannot quantify a mental workload in real-time because it requires a person to perform two different tasks at two different times and compares the performances. Also, performance scores such as accuracy and completion time are not good measures for representing the person's mental workload since mental workload is only one factor used to determine performance.

Second, subjective-rating techniques require the person to rate the difficulty level of a task usually after a given task. This technique can not measure the person's mental workload in real-time, does not objectively quantify mental workloads, and interrupts the main task to record rating.

Third, although physiological measures are good at objectively quantifying mental workload in real-time with no or little interruption of main tasks, most of these physiological measures have other disadvantages. For example, heart rate and respiratory rate measures are not direct in quantifying mental workload. The method requires a model which gives a causal relation between mental workload and those measures. Brain imaging techniques such as fMRI and PET techniques are physically obtrusive to a person, and the devices used for them are too large to equip in a vehicle.

All the previously existing mental workload measures have more than one shortcoming when it comes to quantifying driver's mental workloads. Recently, many researcher groups have started focusing on studying other kinds of physiological measures, such as eye movements, in the hopes of finding better mental workload assessment techniques [2, 3]. Among these physiological measures, vestibule-ocular reflex (VOR) has grabbed the attention of researchers and has been examined to assess its effectiveness in quantifying mental workloads Furman et al [4], Shibata, Obinata and Kajiwara [5], Shibata, Obinata, Kodera and Hamada [6]. This VOR method of quantifying mental workload has seven major advantages over other existing mental workload measures: The VOR method is (1) objective, (2) does not interrupt the main tasks, (3) measurable in real-time, (4) quickly reflects mental workload, (5) accurate, (6) not physically obtrusive, and (7) does not require large equipment. Taking from the previous studies, in this chapter we use the *n*-back tasks to determine if VOR responses could be reasonable measures for quantifying a person's mental workloads.

12.2 Model of Eye Movement

12.2.1 Model of VOR

A block diagram of the proposed eye movement model is shown in Fig. 12.1. We assume in this study that a subject performed a verbal *n*-back task with vibration of the seat and is asked to keep his gaze at a fixed point. Then, VOR is continuously evoked to stabilize his point of gaze against disturbance on the head position during the task. The VOR model which has been proposed by D.M. Merfeld et al. [7] is a mathematical model that represents the dynamics of human VOR responses in three dimensions and is shown in Fig. 12.1(I). This VOR model expresses the interaction of human semicircular canals and statoliths, which sense the head rotation and orientation. Moreover, it contains human internal processing to estimate a predicted value that is the feed-forward

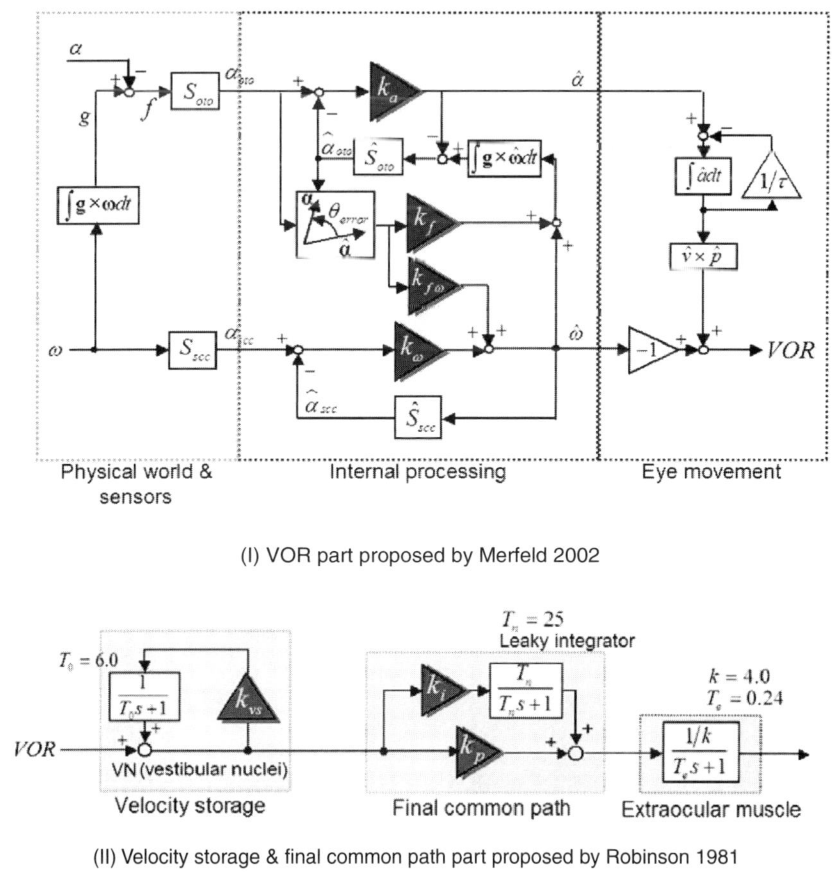

Fig. 12.1 Vetibulo-ocular reflex model

signal to generate motions of eyeball against disturbance in the head position [7]. The angular velocity and linear acceleration of the head in three dimensions are the inputs to this model. Then, it outputs the angular velocity of eyeball in three dimensions. It has been suggested by Robinson [8] that the eventual angle of eyeball, which defines the gaze direction, is not a simple integration of the angular velocity. The transformation from VOR output to the angle is shown in Fig. 12.1(II). The signal is modified through velocity storage, final common path and extraocular muscle. Consequently, we apply this composed VOR model as the eye movement model of a subject.

12.2.2 Model Identification

So as to evaluate mental work load during *n*-back task, we use the eye movement model of each subject. The model must be identified in advance from the experimental data of eye movement which is estimated from rotation of an optokinetic pattern (black–white stripe). The VOR model contains nine gain parameters. These parameters have been determined by minimizing the error between the measured output and the model output. The VOR movements were induced by shaking the subject with moving chair while he is asked to stabilize his eye on the center of screen. The experimental setup and locations of sensors are shown in Fig 12.2.

We have applied the genetic algorithm with local search of gradient method to estimate the parameters which minimize the error norm expressed by

$$J = \sum_{i=1}^{N} \{\theta_{\rm obs}(i) - \theta_{\rm mdl}(i)\}^2, \tag{12.1}$$

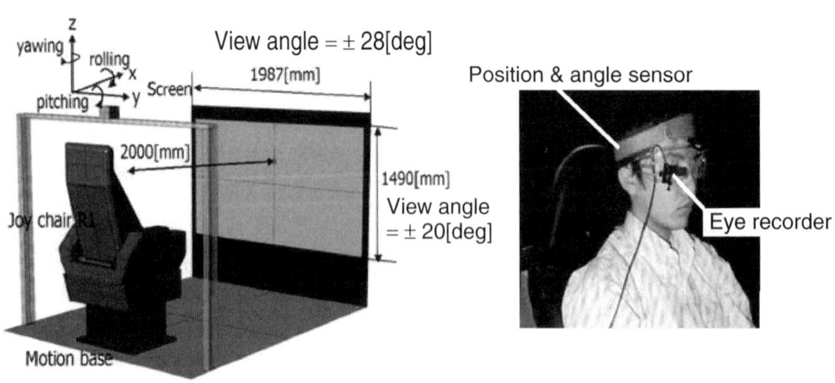

Fig. 12.2 Experimental setup and sensors setting

where θ_{obs} is time series with three dimensions sampled from the observed data of eye movement and θ_{mdl} is time series estimated by the model. Before the identification, the measured eye movement has been treated with outlier removal process that extracts the rapid eye velocity data, which includes blinking and saccade. This is because the dynamics of VOR does not have any relation to blinking and saccade.

The identified model makes it possible to obtain the estimation output of the eye movement. It represents the VOR dynamics of a particular subject who is asked to put his gaze on the center of front screen and is not distracted by any other task. We can use the model to evaluate online the deviation of eye movement when some distraction is given as secondary task.

12.2.3 Identification Results

An example of the identification results is shown in Fig. 12.3. The time responses ((a) (I) and (II)) and the frequency responses of measured eye movements ((b) (I)) and the estimates from the model ((b) (II)) are compared in the figure.

We apply the values for gain parameters as the initial values of search procedure which have been suggested in Merfeld [7] and Robinson [8]. The frequency responses are shown in power spectra and the coherence between the measured eye movements and the predicted values from the identified models. The coherence is defined as

$$\gamma_{xy}^2(f) = \frac{|C_{XY}(f)|^2}{P_{XX}(f)P_{YY}(f)}, \qquad (12.2)$$

where $P_{XX}(f)$ and $P_{YY}(f)$ are power spectra of the measured eye movement and the predicted signal, respectively, and $C_{XY}(f)$ is cross-spectrum between them. The coherence takes unit value when the relation of two signals can be described by a linear differential equation.

We can confirm from Fig. 12.3 that the identification has resulted in a good matching of the model to represent the subject's dynamics of eye movement in both the time responses and the frequency responses. We conducted experiments with three subjects for obtaining the identified models. The results show that the averages of predicted errors in three subjects were from 3 to 6 degrees during eye movements for the range of 20 degrees in horizontal direction and 30 degrees in vertical direction. It should be noted that the identified parameters vary in a certain range in three subjects. This suggests a further research to understand the variation in individuals.

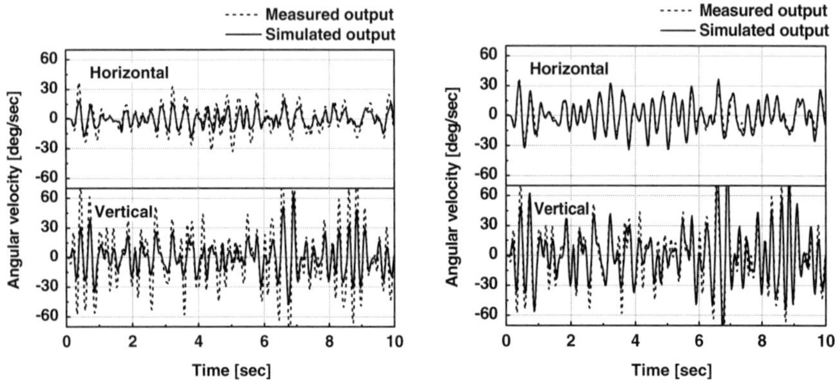

(a) Time Series Data

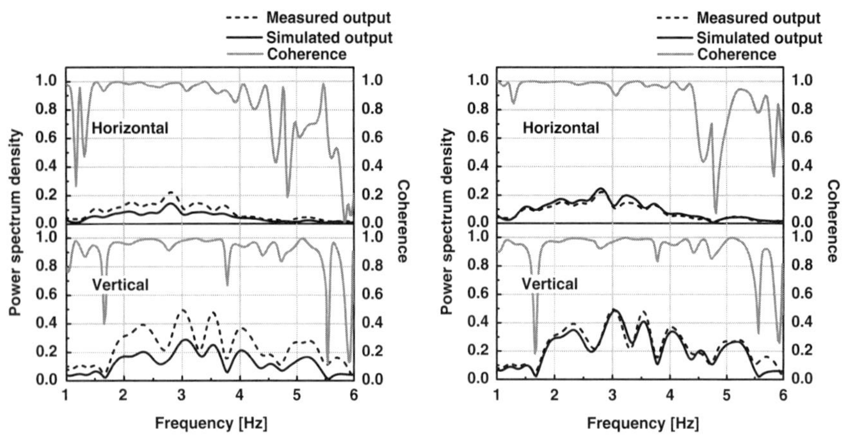

(I) Before Identification (II) After Identification

(b) Frequency Analysis

Fig. 12.3 Example of identification results

12.3 Method Of Experiment

12.3.1 Experiment Procedure

Working memory attracts researcher's attentions as a function of human brain because the function has been visible recently by non-invasive imaging of brain. It is well known that tasks in association with working memory are sensitive to information processing demands [9]. The *n*-back tasks are usually used in working memory research because the *n*-back tasks require the person to maintain and update information at a certain pace. Past research has revealed that during the *n*-back tasks, the activated areas were the frontal association area, temporal association area, and Broca's area [10], [11]. Loads of *n*-back tasks are different by person. However, for any person, a higher-number-back task universally requires more mental workload than a lower-number-back task.

The proposed eye movement model is described with the block diagram in Fig. 12.1. We assume in this study that a subject performed a verbal "*n*-back" task with seat vibration and is asked to keep his gaze at a fixed point of screen. Then, VOR is continuously evoked to stabilize his point of gaze against disturbance on the head position during the task. In the "*n*-back" task, subjects must decide for each verbally presented letter whether it matches the letter presented *n* items back in sequence. The subject was required to respond every 2.5 seconds with push switches as shown in Fig. 12.4. The duration of one trial was 30 seconds. Thus, the subject was required to respond for 12 questions for each trial. Three subjects (ages 20–24, males) participated in the work.

The *n*-back tasks require the subjects for holding/updating information, and decisions based on it. Working memory acts dynamically for such functioning during *n*-back tasks. Several studies on imaging brain functions by MRI have provided the observations that frontal association area, temporal association area, and Broca's area are activated during *n*-back tasks [9], [10].

The experimental procedure is as follows:

1. The subject was asked to sit in moving chair and to keep his gaze at red point of screen.
2. Four types of tasks were given to respond with push switches. Those are simple reaction tasks, 1-back, 2-back, and 3-back tasks. In each trial, the subject was required to respond every 2.5 seconds using switches whether the answer was Yes or No.

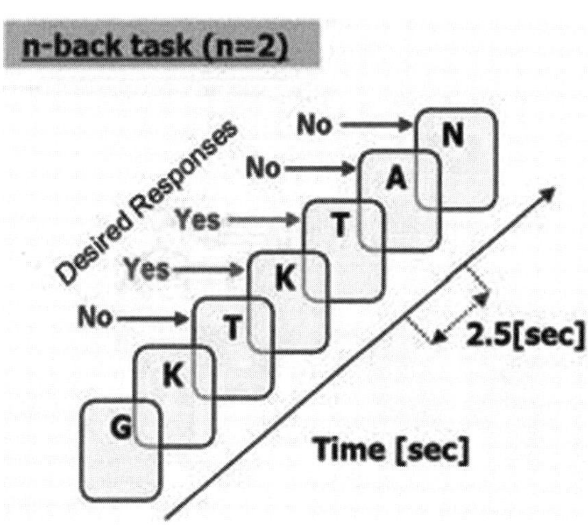

Fig. 12.4 Verbal *n*-back task

3. Three trials were conducted for each type of task. The eye movements and reaction times to every verbal presentations of letter were recorded.

12.3.2 Results of Experiment

The experimental results are shown with the proportion correct (the rate of right answers) and the reaction time: the average and the standard deviation in Fig. 12.5. The reaction times are normalized by the average time of simple reaction task. The proportion correct goes down a little as n increases; on the other hand, the reaction time goes up steadily. These results suggest that the subject takes a certain time for manipulating remembered information in working memory and the time spent increases as the information increases.

The power spectrum and coherence between the measured eye movements and the predicted signals from the identified VOR model are shown in Fig. 12.6. It is noted that the frequency range higher than 4 Hz in the horizontal direction contains less power because of the spectrum of seat vibration. We cannot find out any clear difference in power spectra in all cases. The curve (II) of coherence for 1-back task is similar to the control (I). The coherence of 2-back (III) and 3-back (IV) take lower values in a majority of the frequency range in comparison with the control (I). This means that the dynamics of VOR change in 2-back and 3-back tasks. The average of coherence in n-back tasks over all subjects is shown in Fig. 12.7 in comparison with the control condition and simple reaction task. The results of t-test are also shown in the figure. These results suggest that online evaluation of mental work load is possible by calculating frequency characteristics of eye movements with the reference model under control condition. In addition, we calculated phase shift between the measured eye movement and the predicted signal of the VOR model. Certain phase shifts were observed when n-back tasks were given. In order to complete the analysis on the phase shift, we may have to take the time-varying properties of the signals into consideration.

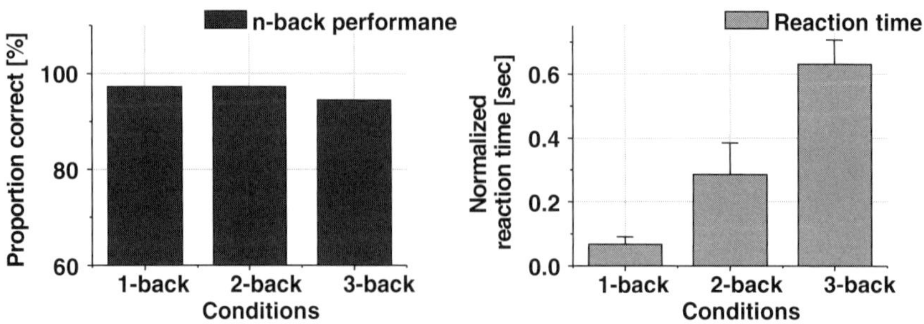

Fig. 12.5 Experimental results

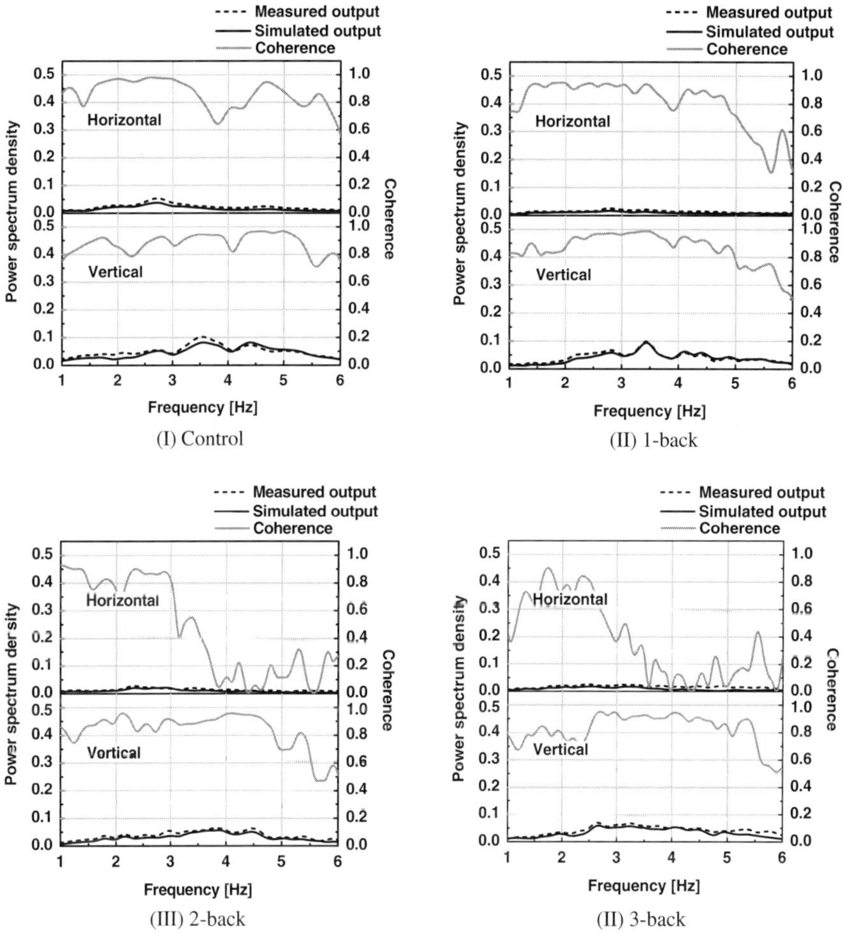

Fig. 12.6 Comparison of coherences and power spectrum in the tasks

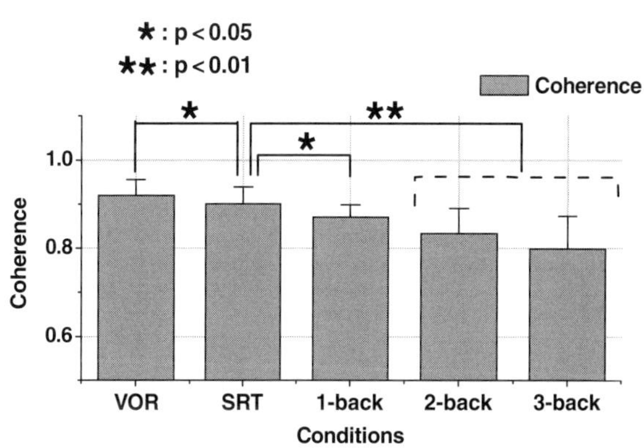

Fig. 12.7 Performance of the influence of working memory on the coherence average (over all subjects)

12.4 Conclusion

This study identified the eye movement model, which represents the dynamics of VOR. It is shown with experimental results of *n*-back tasks that the information processing demand caused by the tasks affects involuntary eye movement for gaze stabilization. In addition, a model-based method has been proposed to quantitatively evaluate the influence of demanded tasks on VOR dynamics. It is shown that the results of evaluation have a relationship to the memory amount of demand. Consequently, our proposed model-based method makes it possible to quantitatively evaluate the effect of demanded tasks. It is worth to mention that the proposed model-based evaluation method can be carried out in real time. It has the possibility to estimate the information processing demand to human operators on the tasks and jobs. It also has the potential to be applied to a driver assistance system or information control system for in-vehicle HMI.

Further research will be devoted to investigating other factors, for example, time-varying characteristics of VOR and the difference of the dynamics among individuals. In addition, further research will aim to establish a method to estimate not only the depth of influence of demanded tasks but also the timing when a subject gets distracted from his primary task. The estimation of timing makes it possible to give the operator a warning. This kind of warning may be useful for driver assistance systems.

Acknowledgments The authors would like to thank Toyota Motor Co. Ltd. for giving the motivation of this study. The work of Professor Kazunori Hase and Mr. Satoru Tokuda on his handling and improving the manuscript are also gratefully acknowledged.

References

1. Stanton, N. A., Salmon, P. M., Walker, G. H., Baber, and C, Jenkins, D. P. (2005) *Human Factors Methods, - A Practical Guide for Engineering and Design*, Ashgate, Burlington, VT.
2. Deubel, H. and Schneider, W. X. (1996): Saccade target selection and object recognition: Evidence for a common attentional mechanism. Vision Research 36:1827–1837.
3. Recanzone, G. H. and Wurtz, R. H. (2000). Effects of attention on MT and MST neuronal activity during pursuit initiation. J. Neurophysiol. 83:777–790.
4. Furman, J. M., et al. (2003): Visual-vestibular stimulation interferes with information processing in young and older humans. Exp. Brain Res. 152:383–392.
5. Shibata, N., Obinata, G., and Kajiwara, Y. (2006): Evaluation of driver distractions by model-based approach –effect of visual distractions-. Review of Automotive Engineering, 27–1:109–115.
6. Shibata, N., Obinata, G., Kodera, and H., Hamada, H. (2006): Evaluating the influence of distractions to drivers based on eye movement model. Proc. FISITA World Automotive Congress, Yokohama, F2006D164.
7. Merfeld, D. M. and Zupan, L. H. (2002): Neural processing of gravitoinertial cues in humans. Modeling tilt and translation responses. J. Neurophysiol. 87:819–833.
8. Robinson, D. A. (1981): The use of control system analysis in the neurophysiology of eye movements. Ann. Rev. Neuroscience 4:463–503.

9. Marsha, C. L., et al. (2000): A source activation theory of working memory: cross-task prediction of performance in ACT-R. Journal of Cognitive System Research 1:99–118.
10. Braver, T. S., et al. (1997): A parametric study of prefrontal cortex involvement in human working memory. Neuroimage 5:49–62.
11. Cohen, J. D., et al. (1994): Activation of prefrontal cortex in a non-spatial working memory task with functional MRI. Human Brain Mapping 1:293–304.

13
Estimation of Acoustic Microphone Vocal Tract Parameters from Throat Microphone Recordings

Ülkü Çağrı Akargün and Engin Erzin

Abstract Joint processing of throat and acoustic microphone recordings has been an attractive tool for robust speech processing. As the throat microphones record the acoustic sounds in the form of vibrations from skin-attached sensors, they are more robust and highly correlated with the acoustic speech signal. We investigate the correlation of throat and acoustic microphone recordings. We propose a hidden Markov model (HMM)-based structure to build a mapping between throat and acoustic microphone vocal track parameters. The HMM-based estimator can be used to estimate clean acoustic speech features from noisy acoustic and throat microphone recordings. Experimental results on acoustic speech feature estimation are provided.

Keywords Robust speech processing · Acoustic and throat microphone recordings · HMM-based correlation model

13.1 Introduction

Speech recognition has been a target for natural man–machine communication interface. However, robustness problems under varying environmental conditions are still among the main barriers to the wide use of speech recognition, since environmental robustness degrades speech recognition performance significantly for many application targets. In the last two decades, many research articles address robustness issues in speech recognition under varying environmental conditions. Some of the mainstream robustness studies include speech enhancement, cepstral mean subtraction, model adaptation, where these studies target to increase recognition performance under adverse conditions. Beside these unimodal (speech-only) approaches, recently multimodal approaches try

E. Erzin (✉)
Multimedia, Vision and Graphics Laboratory, Koç University, Sarıyer, İstanbul, 34450, Turkey
e-mail: eerzin@ku.edu.tr

K. Takeda et al. (eds.), *In-Vehicle Corpus and Signal Processing for Driver Behavior*, DOI 10.1007/978-0-387-79582-9_13,

to benefit from robust modalities, such as the use of lip movements within audio–visual speech recognition [1]. Multimodal approaches are beneficial when they include environment-independent but speech-correlated modalities. In this study, we as well develop a multimodal speech signal processing platform, which targets to investigate joint analysis of throat and acoustic microphone (TAM) recordings.

The multimodal approaches became increasingly attractive in the last decade. Among these efforts the joint processing of TAM recordings gained momentum in the last couple of years. As the throat microphones record the acoustic sounds in the form of vibrations through skin-attached sensors at the throat, these recordings are more robust than acoustic microphone recordings to environmental conditions. However, they represent a lower bandwidth speech signal content compared to acoustic microphone recordings. Since the throat microphone recordings are more robust and highly correlated with the acoustic speech signal, they become attractive candidates for robust speech-processing applications.

One of the early works uses the voice vibrations recorded from throat in speech recognition applications; the throat and noisy acoustic microphone speech signals are linearly combined yielding a robust estimate of the clean speech signal [2]. Later, Graciarena et al. [3] proposed combining the acoustic microphone and the throat microphone.

The acoustic microphone clean speech feature vectors are estimated using the probabilistic optimum filter (POF) mapping, which is a piecewise linear transformation of noisy feature space into the clean feature space [4]. A device that combines a close-talk and a bone-conductive microphone is proposed by Microsoft research group [5, 6] for speech detection using a moving-window histogram. Clean speech is estimated from the bone sensor signals in [5]. In [6], the authors aim to learn the mapping from the bone sensor to the close-talk microphone. In order to reconstruct the clean speech signals, the predicted speech from the bone sensor is fused with the noisy close-talk speech. In another combined acoustic and throat microphone processing approach [7], the speech recorded from throat and acoustic channels is processed by parallel speech recognition systems and later a decision fusion yields a more robust recognition to background noise. In [8] a different sensor, non-audible murmur (NAM) device, which can detect very quietly uttered speech, is attached behind the ear. Speech recognition through NAM microphone increases robustness while availing privacy in human–machine communication.

In this chapter, we investigate the correlation between throat and acoustic microphone recordings to build a mapping between throat and acoustic microphone vocal track parameters. The correlation model between throat and acoustic microphone vocal track models and the proposed acoustic vocal tract estimation scheme from the throat microphone recordings are presented in the following section. Later, we present experimental evaluations of the proposed estimation scheme over a parallel TAM corpus.

13.2 Acoustic–Throat Correlation Model

Throat microphone recordings are highly correlated with acoustic microphone signal at low frequencies. A sample parallel spectrogram is given in Fig. 13.1. Hence, we expect a certain degree of correlation between spectral features of the acoustic and throat microphone recordings. In this study, we focused on the analysis of this correlation over the source–filter model. Let us represent the acoustic and throat microphone recordings by $y(t)$ and $b(t)$ and the corresponding source–filter model as

$$Y(w) = E(w)H(w) + N(w) = E_a(w)H_a(w), \tag{13.1}$$

$$B(w) = E(w)H(w)M(w) = E_t(w)H_t(w). \tag{13.2}$$

Here, $e(t)$ is the excitation signal, $h(t)$ is the vocal track filter impulse response, $n(t)$ is the possible additive environmental noise and $m(t)$ is the filter impulse response that forms the vocal tract in throat microphone recordings. We can obtain source $E_{t,a}(w)$ and vocal tract filter $H_{t,a}(w)$ using source–filter analysis of TAM recordings. Vocal tract filter, $H_t(w)$, that corresponds to throat microphone recordings is a low-pass version of the actual vocal tract, $H(w)$, in this model. The throat microphone recordings are more robust to environmental noise than acoustic speech signals, hence they are a good representation of the low-pass nature of the actual vocal tract. We can obtain an estimate of the clean acoustic vocal track parameters from the noisy vocal track filter $H_a(w)$ and the throat microphone vocal track filter $H_t(w)$. Let us define this estimator as

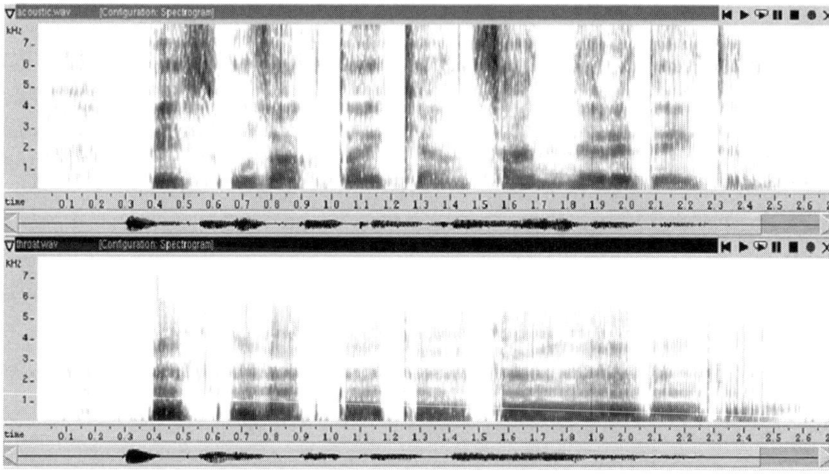

Fig. 13.1 A sample of parallel acoustic and throat microphone spectrograms

$$\hat{H}(w,t_n) = \Phi(H_{\mathrm{a}}(w,t_n-\Delta \le t \le t_n+\Delta), H_{\mathrm{t}}(w,t_n-\Delta \le t \le t_n+\Delta) \quad (13.3)$$

where the function $\Phi(.,.)$ estimates clean vocal tract filter from the noisy acoustic and throat vocal tract filters in the Δ neighborhood of the time instant t_n. The synchronous clean TAM recordings are processed with source–filter analysis and line spectral pairs (LSPs), representing the vocal tract filter, are extracted synchronously for acoustic and throat microphones. Let us define the LSP vectors for acoustic and throat microphone channels as L_k^{a} and L_k^{t} for the kth frame, respectively. The L_k^{a} represents the acoustic microphone filter models $H_{\mathrm{a}}(w)$ and $H(w)$, which are equal to each other for the clean environment recordings. Likewise, L_k^{t} vector represents the throat microphone filter model $H_{\mathrm{t}}(w)$ for the kth frame. Joint synchronous filter parameters can be represented as $L_k^{\mathrm{at}} = [L_k^{\mathrm{aT}}, L_k^{\mathrm{tT}}]^{\mathrm{T}}$ for the kth frame.

We evaluate two different approaches to estimate acoustic microphone filter models from throat microphone filter models. The structure of the general estimator in Eq. (13.3) is modified to estimate the acoustic filter from the throat vocal filter as follows:

$$\hat{L}_k^{\mathrm{a}} = \Phi(L_{k-\Delta}^{\mathrm{t}}, L_{k-\Delta+1}^{\mathrm{t}}, ..., L_k^{\mathrm{t}}, ..., L_{k+\Delta-1}^{\mathrm{t}}, L_{k+\Delta}^{\mathrm{t}}). \quad (13.4)$$

This estimator is realized using two different approaches: (1) the commonly used vector quantization approach and (2) the hidden Markov model (HMM)-based approach. The estimators and their performances are presented in the following sections.

13.2.1 Vector Quantization-Based Estimator

A vector quantizer can be designed to jointly quantize the synchronous TAM vocal track filter parameters. This quantizer can be used to estimate acoustic filter model from the throat filter model. Let the joint vector quantizer L_{QB}^{at} be designed over the multi-stream filter parameter vectors L_k^{at} with 2^B levels using the generalized Lloyd I (also known as the Linde–Buzo–Gray (LBG) training algorithm [9]. Hence each element of L_{QB}^{at} quantizer is a joint acoustic and throat microphone filter parameter vector. L_{QB}^{at} vector quantizer can be split into two conjugate vector quantizers, L_{QB}^{a} and L_{QB}^{t}, which will represent acoustic and throat channels, respectively. We can quantize any throat filter parameter vector L_k^{t} for the kth frame:

$$\begin{aligned} \overline{L}_k^{\mathrm{t}} &= \arg\min_{C\, L_{QB}^{\mathrm{t}}(i)\forall i} ||L_k^{\mathrm{t}} - L_{QB}^{\mathrm{t}}(i)|| \\ I_k^{\mathrm{t}} &= \arg\min_{0 \le i < 2^B} ||L_k^{\mathrm{t}} - L_{QB}^{\mathrm{t}}(i)||. \end{aligned} \quad (13.5)$$

Here, $\overline{L}_k^{\mathrm{t}}$ is the quantized throat filter parameter vector and I_k^{t} is the index of the quantized vector. We can estimate the acoustic filter parameter vector using the quantized throat filter parameter vector $\overline{L}_k^{\mathrm{t}}$ and the conjugate vector quantizer L_{QB}^{a} as

$$\hat{L}_k^{\mathrm{a}} = L_{QB}^{\mathrm{a}}(I_k^{\mathrm{t}}). \tag{13.6}$$

The estimation error between the estimate $\hat{L}_k^{\mathrm{a}}$ and the original L_k^a parameter vectors can be computed as the logarithmic spectral distortion between the estimated and original spectrums $\hat{H}_{\mathrm{a}}(w)$ and $H_{\mathrm{a}}(w)$ as follows:

$$\mathrm{d}(H_{\mathrm{a}}(w), \hat{H}_{\mathrm{a}}(w)) = \frac{1}{2\pi} \int_{-\pi}^{\pi} [10 \log \frac{1}{|H_{\mathrm{a}}(w)|^2} - 10 \log \frac{1}{|\hat{H}_{\mathrm{a}}(w)|^2}]^2 \mathrm{d}w \tag{13.7}$$

13.2.2 Hidden Markov Model-Based Estimator

Recently, we developed an unsupervised segmentation-based multimodal correlation analysis framework [10]. We adapt this hidden Markov model (HMM)-based unsupervised multi-stream analysis framework to determine a correlation model between acoustic and throat microphone recordings. The HMM-based unsupervised classifier is used to jointly segment temporal spectral features of the TAM recordings. The joint temporal feature patterns are used to form a correlation model between the throat and acoustic streams. The multi-stream unsupervised segmentation defines recurrent sub-phonetic patterns of the joint TAM recordings. The acoustic and throat microphone feature streams L^{a} and L^{t} are jointly used to train the HMM structure, Λ^{at}, which captures recurrent phonetic segments. The HMM structure Λ^{at}, which is used for unsupervised temporal segmentation, is composed of M parallel left-to-right HMMs, $\{\lambda_1^{\mathrm{at}}, \lambda_2^{\mathrm{at}}, \ldots, \lambda_M^{\mathrm{at}}\}$, where each λ_m^{at} is composed of N states, $\{s_{m,1}, s_{m,2}, \ldots, s_{m,N}\}$, as shown in Fig. 13.2. Given the multimodal feature sequence, $L^{\mathrm{at}} = \{L_1^{\mathrm{at}}, L_2^{\mathrm{at}}, ..., L_K^{\mathrm{at}}\}$, L_k^{at} denotes the joint feature vector at frame k. The segmentation of the feature sequence is performed using Viterbi decoding to maximize the probability of model match. The Viterbi decoding yields a state sequence $q^{\mathrm{at}} = \{q_1^{\mathrm{at}}, q_2^{\mathrm{at}}, ..., q_K^{\mathrm{at}}\}$ associated with the feature sequence L^{at}.

The proposed throat-driven acoustic feature estimation process takes throat microphone recordings as input and produces a sequence of robust acoustic features. The multi-stream HMM, which was obtained in the joint analysis of TAM recordings, is split into throat only, Λ^{t}, and acoustic only, Λ^{a}, models. Given the throat model Λ^{t} and throat microphone recordings, the flow of the estimation is described with the following steps:

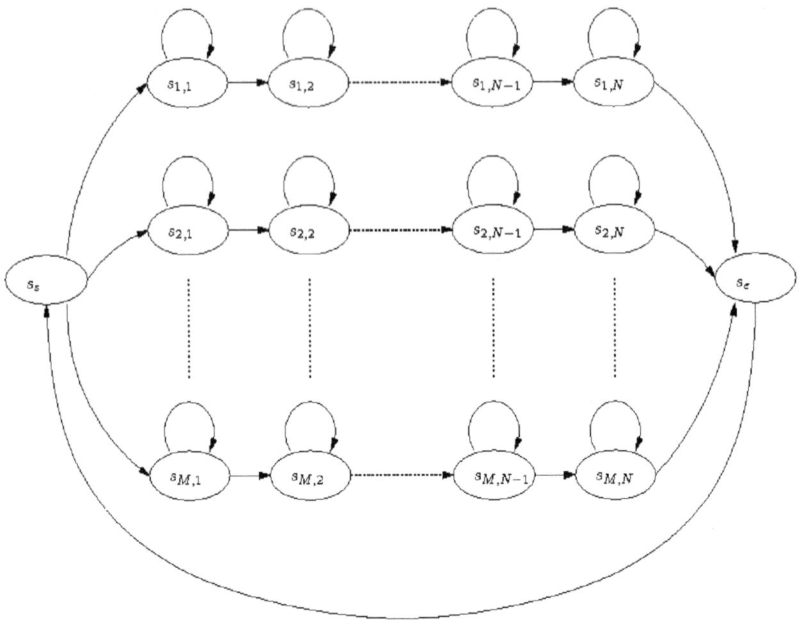

Fig. 13.2 Hidden Markov model-based unsupervised classifier

1. The throat features, L^{t}, are extracted from the input throat microphone recordings.
2. Temporal segmentation of throat feature sequence L^{t} is performed using the HMM model Λ^{t} to extract the temporal sub-phone patterns with a state sequence q^{t}.
3. Then, the acoustic feature sequence $\hat{L}^{\mathrm{a}}$ is estimated by maximizing the probability of observing acoustic feature sequence on the state sequence path q^{t} over the acoustic HMM model Λ^{a}.

13.3 Experimental Results

We built a synchronous acoustic and throat microphone database, which consists of 400 sentence recordings from a single subject under clean conditions. In our experimental studies we split this database into two equal parts to perform training and testing of the vector quantizer and HMM-based estimators. The LSP parameters are calculated using 16th order linear prediction filters over a window of size 20 ms for every 10 ms frames.

The performance of the vector quantizer-based estimator is analyzed for varying codebook sizes. The average log-spectral distortion values within train and test sets are given in Fig. 13.3. As expected, spectral distortion tends to decrease as the dimension of the vector quantizer increases for the training set.

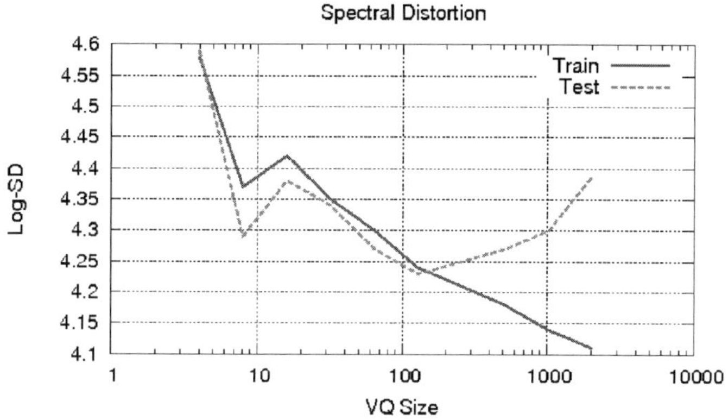

Fig. 13.3 The average log-spectral distortion performance of the vector quantizer-based estimator on training and test sets

As for the test data, first the spectral distortion tends to decrease and then increase. It is observed that the best estimation is obtained with a 128 codebook size.

A similar performance analysis is done for the HMM-based estimator. The average log-spectral distortion performances for varying number of classes with single and two Gaussian mixtures are given in Figs. 13.4 and 13.5, respectively. The best performances are obtained with 37-class HMM structure with single Gaussian mixture and 33-class HMM structure with two Gaussian mixtures. Note that there is equivalence between vector quantizer and HMM-based estimators in terms of total number of clusters. The HMM-based estimator has 4 states in a single branch and 37/33 total branches, on the other hand vector quantizer-based estimator has 128 codebooks. However, HMM-based estimator captures the temporal changes, and with this property, it is observed that the

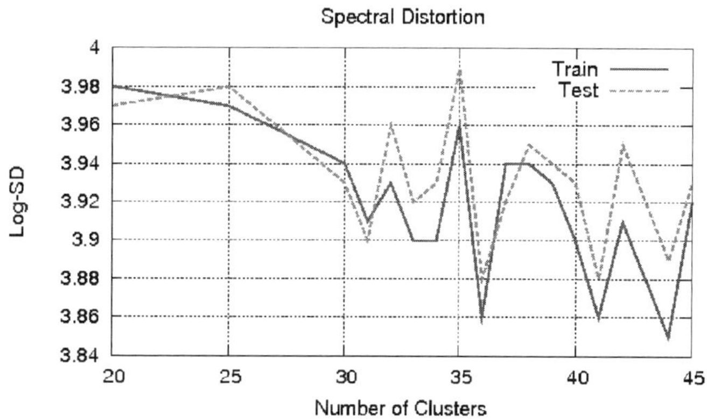

Fig. 13.4 The average log-spectral distortion performance of the HMM-based estimator with single Gaussian mixture on training and test sets over varying number of classes

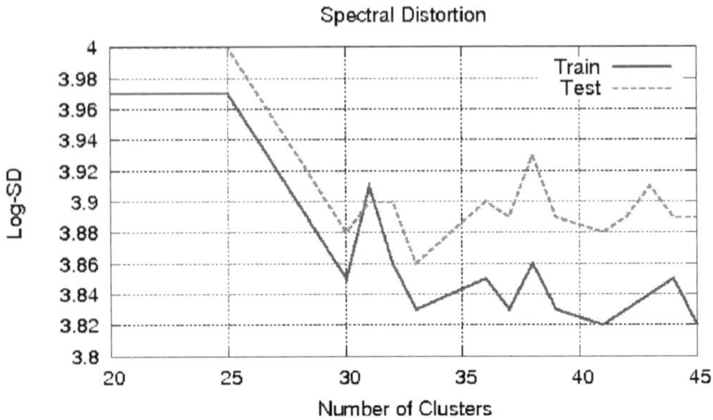

Fig. 13.5 The average log-spectral distortion performance of the HMM-based estimator with two Gaussian mixtures on training and test sets over varying number of classes

HMM-based estimator attains lower log-spectral distortion than the vector quantizer-based estimator.

13.4 Conclusions

In this chapter, we have focused on the analysis of the correlation between the throat and acoustic microphone recordings. Vector quantization and HMM-based estimators are examined under clean environment recordings to estimate acoustic filter parameters from throat filter parameters. We observed that the average log-spectral distortion values for the HMM-based estimator is better compared to the vector quantization-based estimator. The use of HMM-based correlation analysis of the TAM recordings under adverse conditions for robust speech recognition is currently under investigation.

Acknowledgments The authors would like to acknowledge the support of TUBITAK under project EEEAG-104E176, the state planning organization of Turkey (DPT) under Drive-Safe project, and NEDO (New Energy and Development Organization) of Japan.

References

1. Cetingul HE, Erzin E, Yemez Y, Tekalp AM (2006) Multimodal speaker/speech recognition using lip motion, lip texture and audio. Signal Processing, Special Section: Multimodal Human-Computer Interfaces, 86(12):3549–3558.
2. Roucos S, Viswanathan V, Henry C, Schwartz R (1986) Word recognition using multisensor speech input in high ambient noise. IEEE International Conference on Acoustics, Speech, and Signal Processing, 11:737–740.
3. Graciarena M, Franco H, Sonmez K, Bratt H (2003) Combining standard and throat microphones for robust speech recognition. IEEE Signal Processing Letters, 10(3):72–74.

4. Neumeyer L, Weintraub M (1994) Probabilistic optimum filtering for robust speech recognition. IEEE International Conference on Acoustics, Speech, and Signal Processing, I:417–420.
5. Zheng Y, Liu Z, Zhang Z, Sinclair M, Droppo J, Deng L, Acero A, Huang X (2003) Air- and bone-conductive integrated microphones for robust speech detection and enhancement. Proceedings of ASRU 2003, U.S. Virgin Islands, December 2003.
6. Zhang Z, Liu Z, Sinclair M, Acero A, Deng L, Droppo J, Huang X, Zheng Y (2004) Multi-sensory microphones for robust speech detection, enhancement, and recognition. IEEE International Conference on Acoustics, Speech, and Signal Processing, Montreal, May 2004.
7. Dupont S, Ris C (2004) Combined use of close-talk and throat microphones for improved speech recognition under non-stationary background noise. Proceedings of Robust 2004 (Workshop (ITRW) on Robustness Issues in Conversational Interaction), Norwich, August 2004.
8. Heracleous P, Nakajima Y, Lee A, Saruwatari H, Shikano K (2004) Non-audible murmur (NAM) speech recognition using a stethoscopic NAM microphone. Proceedings of INTERSPEECH-2004, pp. 1469–1472.
9. Linde Y, Buzo A, Gray RM (1980) An algorithm for vector quantizer design. IEEE Transactions on Communications, January:702–710.
10. Sargin ME, Erzin E, Yemez Y, Tekalp AM, Erdem AT, Erdem C, Ozkan M (2007) Prosody-driven head-gesture animation. IEEE International Conference on Acoustics, Speech and Signal Processing, II:677–680, April 2007.

14
Cross-Probability Model Based on Gmm for Feature Vector Normalization

Luis Buera, Antonio Miguel, Eduardo Lleida, Alfonso Ortega, and Óscar Saz

Abstract In order to develop a robust man–machine interface based on speech for cars, multi-environment model-based linear normalization, MEMLIN, was presented earlier and it was proved to be effective to compensate environment mismatch. MEMLIN is an empirical feature vector normalization technique which models clean and noisy spaces with Gaussian mixture models, GMMs; and the probability of the clean model Gaussian, given the noisy model one and the noisy feature vector (cross-probability model), is a critical point. In previous works the cross-probability model was approximated as time independent in a training process. However, in this chapter, an estimation based on GMM is considered for MEMLIN. Some experiments with SpeechDat Car and Aurora2 databases were carried out in order to study the performance of the proposed estimation of the cross-probability model, obtaining important improvements: 75.53 and 62.49% of mean improvement in word error rate, WER, for MEMLIN with SpeechDat Car and a reduced set of Aurora2 database, respectively (82.86 and 67.52% if time-independent cross-probability model is applied). Although the behaviour of the technique is satisfactory, using clean acoustic models in decoding produces a mismatch because the normalization is not perfect. So, retraining acoustic models in the normalized space is proposed, reaching 97.27% of mean improvement with SpeechDat Car database.

Keywords Robust automatic speech recognition · Feature vector normalization · GMM · MEMLIN · Cross-probability model

14.1 Introduction

Since cars are more and more considered as business offices, drivers need a safe way to communicate and interact with either other people or machines. For safety reason, traditional visual and tactile man–machine interfaces, such

---

L. Buera (✉)
Aragon Institute of Engineering Research (I3A), University of Zaragoza, Zaragoza, Spain

K. Takeda et al. (eds.), *In-Vehicle Corpus and Signal Processing for Driver Behavior*, DOI 10.1007/978-0-387-79582-9_14,

as displays, buttons, and knobs are not satisfactory but speech, as the most convenient and natural way of communications is an appropriate and complementary solution which can reduce distractions. Hence, automatic speech recognition (ASR) provides safety and comfort, and it is possible to follow the philosophy "Eyes on the road and hands on the steering wheel", which should drive every in-vehicle system design. The problem of robust ASR in car environments has attracted much attention in the recent years and a new market demands for systems which allow the driver to control non-critical devices or tasks like phone dialling, RDs-tuner, air conditioner, satellite navigation systems, remote information services access, Web browsing. For this purpose, robustness in challenging car environment still needs to be improved.

When training and testing acoustic conditions differ, the accuracy of ASR systems rapidly degrades. To compensate for this mismatch, robustness techniques have been developed along the following two main lines of research: acoustic model adaptation methods and feature vector adaptation/normalization methods. Also, hybrid solutions, which are effective under certain conditions, can be generated by combining both kinds of techniques [11]. In general, acoustic model adaptation methods produce the best results [10] because they can model the uncertainty caused by the noise statistics. However, these methods require more data and computing time than do feature vector adaptation/ normalization methods, which do not produce as good results but provide more online solutions. So, finally, the choice of a robustness technique depends on the characteristics of the application in each situation.

Feature vector adaptation/normalization methods fall into one of three main classes [12]: high-pass filtering, which contains very simple methods such cepstral mean normalization, CMN; model-based techniques, which assumes a structural model of environmental degradation; and empirical compensation, which uses direct cepstral comparisons. In any case, and independently of the class, some algorithms assume a prior probability density function (pdf) for the estimation variable. In those cases, a Bayesian estimator can be used to estimate the clean feature vector. The most commonly used criterion is to minimize the mean square error (MSE), and the optimal estimator for this criterion, minimum mean square error (MMSE), is the mean of the posterior pdf. Methods, such as *s*tereo-based *p*iecewise *li*near *c*ompensation for *e*nvironments (SPLICE) [4] or *m*ulti-*e*nvironment *m*odel-based *li*near *n*ormalization (MEMLIN) [1] use the MMSE estimator to compute the estimated clean feature vector.

Previous works [1] show that MEMLIN is effective to compensate the effects of dynamic and adverse car conditions. MEMLIN is an empirical feature vector normalization technique based on stereo data and the MMSE estimator. MEMLIN splits the noisy space into several basic environments and each of them and clean feature space are modelled using GMMs. Therefore, a bias vector transformation is associated with each pair of Gaussians from the clean and the noisy basic environment spaces. A critical point in MEMLIN is the estimation of the cross-probability model (the probability of the clean model Gaussian, given the noisy model one, and the noisy feature vector). In [1], a time-independent solution is considered to compute this probability, but this work focuses on a different solution [2], which consists of modelling the noisy feature vectors associated to each pair of Gaussians from the clean and the noisy basic environment spaces with a GMM. Furthermore, adapting acoustic models to the normalized space is proposed to reduce the mismatch between compensated feature vectors and clean acoustic models.

This chapter is organized as follows: In Section 14.2, an overview of MEMLIN is detailed. In Section 14.3, some experiments are presented to show the importance of the cross-probability model estimation. The GMM-based solution considered to compute the cross-probability model is explained in Section 14.4. The acoustic model retrained is explained in Section 14.5. The results with Spanish SpeechDat Car [9] and Aurora2 [6] databases are included in Section 14.6. Finally, the conclusions are presented in Section 14.7.

14.2 Memlin Overview

MEMLIN is an empirical feature vector normalization technique which uses stereo data in order to estimate the different compensation linear transformations in a previous training process. The clean feature space is modelled as a mixture of Gaussians. The noisy space is split into several basic acoustic environments and each one is modelled as a mixture of Gaussians. The linear transformations are estimated for all basic environments between a clean Gaussian and a noisy Gaussian. The realization scheme for MEMLIN used in this study is shown in Fig. 14.1.

14.2.1 MEMLIN Approximations

- Clean feature vectors, x_t, are modelled using a GMM of C components:

$$p(x_t) = \sum_{s_x=1}^{C} p(x_t|s_x)p(s_x) \tag{14.1}$$

$$p(x_t|s_x) = N(x_t; \mu_{s_x}, \Sigma_{s_x}), \tag{14.2}$$

where t is the time index and μ_{s_x},Σ_{s_x}, and $p(S_x)$ are the mean vector, the diagonal covariance matrix, and the a priori probability associated with the clean model Gaussian s_x.

- Noisy space is split into several basic environments, e, and the noisy feature vectors, y_t, are modelled as a GMM of C' components for each basic

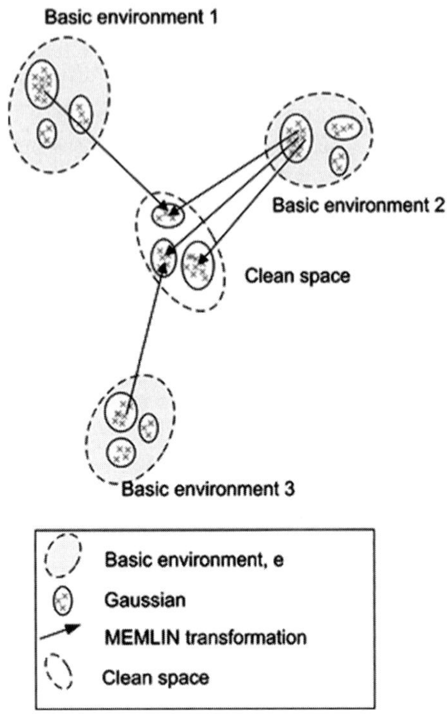

Fig. 14.1 MEMLIN structure

environment (assuming that all the basic environments are modelled with the same number of components):

$$p(y_t) = \sum_{s_y^e=1}^{C'} p\left(y_t \middle| s_y^e\right) p\left(s_y^e\right) \tag{14.3}$$

$$p\left(y_t \middle| s_y^e\right) = N\left(y_t; \mu_{s_y^e}, \Sigma_{s_y^e}\right), \tag{14.4}$$

where s_y^e denotes the corresponding Gaussian of the noisy model for the e basic environment; $\mu_{s_y^e}$, $\Sigma_{s_y^e}$, and $p\left(s_y^e\right)$ are the mean vector, the diagonal covariance matrix, and the *a priori* probability associated with s_y^e, respectively.

- Clean feature vectors can be approximated as a linear function, ψ, of the noisy feature vector which depends on the basic environments and the clean and noisy model Gaussians: $\psi\left(y_y, s_x, s_y^e\right) = y_t - r_{s_x, s_y^e}$, where r_{s_x, s_y^e} is the bias vector transformation between noisy and clean feature vectors for each pair of Gaussians, S_x and s_y^e.

14.2.2 MEMLIN Enhancement

With those approximations, MEMLIN transforms the MMSE estimation expression, *bbb*, into

$$\hat{x}_t = y_t - \sum_{e}\sum_{s_y^e}\sum_{s_x} r_{s_x,s_y^e} p(e|y_t) p\left(s_y^e|y_t, e\right) p\left(s_x\middle|y_t, e, s_y^e\right). \tag{14.5}$$

Here $p(e|y_t)$ is the *a posteriori* probability of the basic environment; $p\left(s_y^e|y_t, e\right)$ is the *a posteriori* probability of the noisy model Gaussian, s_y^e, given the feature vector and the basic environment. To estimate those terms ($p(e|y_t)$ and $p\left(s_y^e|y_t, e\right)$), expressions (14.3) and (14.4) are applied. (described with detail in [1]). Finally, the cross-probability model, $p\left(s_x\middle|y_t, e, s_y^e\right)$, is the probability of the clean model Gaussian, s_x, given the noisy feature vector, the basic environment, and the noisy model Gaussian. The cross-probability model can be estimated avoiding the time dependence given by the noisy feature vector in a training phase using stereo data for each basic environment:

$$\left(X_e^{\mathrm{Tr}}, Y_e^{\mathrm{Tr}}\right) - \left(x_1^{\mathrm{Tr},e}, y_1^{\mathrm{Tr},e}\right); \ldots; \left(x_{t_e}^{\mathrm{Tr},e}, y_{t_e}^{\mathrm{Tr},e}\right); \ldots; \left(x_{T_e}^{\mathrm{Tr},e}, y_{T_e}^{\mathrm{Tr},e}\right),$$

with $t_e \in [1, T_e]$ [1] as

$$p\left(s_x\middle|y_t, e, s_y^e\right) \approx p\left(s_x\middle|e, s_y^e\right) = \frac{\sum_{t_e} p\left(x_{t_e}^{\mathrm{Tr},e}|s_x\right) p\left(y_{t_e}^{\mathrm{Tr},e}\middle|s_y^e\right) p(s_x) p\left(s_y^e\right)}{\sum_{s_x}\sum_{t_e} p\left(x_{t_e}^{\mathrm{Tr},e}|s_x\right) p\left(y_{t_e}^{\mathrm{Tr},e}\middle|s_y^e\right) p(s_x) p\left(s_y^e\right)} \tag{14.6}$$

On the other hand, the bias vector transformation, r_{s_x,s_y^e}, is also computed using the stereo data in the previous training phase [1].

14.3 Cross-Probability Model Performance

To study the performance of the cross-probability model in a qualitative way, the histograms and log-scattergrams between the first Mel frequency cepstral coefficients (MFCCs) in non-silence frames for different signals are depicted in Fig. 14.2.

Figure 14.2.a, which represents the clean and noisy feature coefficients in real car conditions, shows the effects of car noise. The pdf of clean first MFCCs is clearly affected (Fig. 14.2.a.1), and the uncertainty is increased (Fig.14.2.a.2).

In Fig. 14.2.b and 14.2.c, clean and normalized coefficients with MEMLIN (128 Gaussians are considered to model the clean and basic environment

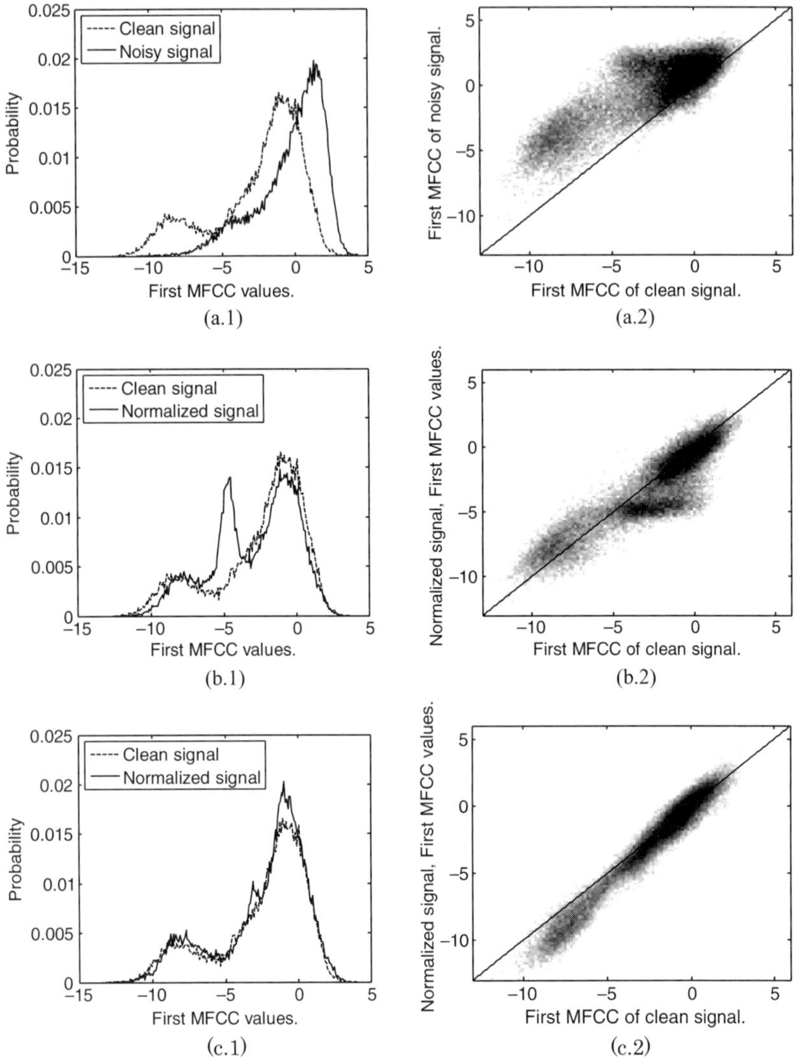

Fig. 14.2 Log-scattergrams and histograms between the first MFCC in non-silence frames for different signals. The diagonal line in the log-scattergrams represents the function $x = y$

spaces) are represented. The pdf of normalized first MFCCs has been approximated to the clean signal one (Fig. 14.2.b.1), and the uncertainty has been reduced (Fig. 14.2.b.2). The peak that appears in Fig. 14.2.b.1 is due to the transformation of noisy feature vectors toward the clean silence.

Finally, Fig. 14.2.c represents clean and normalized coefficients with MEMLIN when the cross-probability model is computed with the corresponding clean feature vector as (14.7); 128 Gaussians are used to model the different spaces. In this case the pdf of the normalized signal is almost the same as the clean one (Fig. 14.2.c.1) and the uncertainty is drastically reduced (Fig. 14.2.c.2). Furthermore, the WER results in this case are almost the same that we would obtain with clean signal. These results verify the importance of a good estimation of the cross-probability model in MEMLIN algorithm:

$$p\left(s_x \middle| y_t, e, s_y^e\right) \approx \frac{p(s_x)p(x_t|s_x)}{\sum\limits_{s_x} p(s_x)p(x_t|s_x)} \qquad (14.7)$$

14.4 Cross-Probability Model Based on GMM

To improve the time-independent cross-probability model (14.6), we propose to model the noisy feature vectors associated with a pair of Gaussians (s_x and s_y^e) with a GMM of C'' components (assuming that the noisy feature vectors are modelled with the same number of Gaussian as for all pairs s_x and s_y^e). Since the estimation of the corresponding GMMs for each basic environment can be considered independent, they are not indexed to simplify the notation. Hence we present a model of the noisy feature vectors associated with the pair of Gaussians s_x and s_y

$$p(y_t|s_x, s_y) = \sum_{s'_y=1}^{C'} p\left(y_t \middle| s_x, s_y, s'_y\right) p\left(s'_y \middle| s_x, s_y\right) \qquad (14.8)$$

$$p\left(y_t \middle| s_x, s_y, s'_y\right) = N\left(y_t; \mu_{s_x,s_y,s'_y}, \Sigma_{s_x,s_y,s'_y}\right). \qquad (14.9)$$

Here μ_{s_x,s_y,s'_y}, Σ_{s_x,s_y,s'_y}, and $p\left(s'_y \middle| s_x, s_y\right)$ are the mean vector, the diagonal covariance matrix, and the a priori probability associated with s'_y Gaussian of the cross-probability GMM associated with s_x and s_y. To train these three parameters, the EM algorithm [3] is applied. Let a set of clean and noisy stereo data available to learn the corresponding cross-probability GMM parameters

$$(X, Y) = (x_1, y_1); \ldots; (x_n, y_n); \ldots; (x_N, y_N)$$

Each y_n can be seen as an incomplete component-labelled frame, which is completed by two indicator vectors. The first one is $w_n \in \{0, 1\}^{C\prime}$, with 1 in the position corresponding to the s_y Gaussian which generates y_n and zeros elsewhere, $W = \{w_1, \ldots, w_N\}$. The second indicator vector is $z_n \in \{0, 1\}^{C''}$, with 1 in the position corresponding to the s'_y Gaussian of the cross-probability GMM which generates y_n and zeros elsewhere, $Z = \{z_1, \ldots, z_N\}$. Each x_n can be seen also as an incomplete component-labelled frame, which is completed by one indicator vector: $v_n \in \{0, 1\}^C$, with 1 in the position corresponding to the s_x

Gaussian which generates x_n and zeros elsewhere, $V = \{v_1, \ldots, v_N\}$. The indicator vectors are called missing data, too. So, the complete data pdf is

$$p(x, y, v, w, z) \approx p(v, w)p(x|v, w)p(v, w, z)p(y|v, w, z). \tag{14.10}$$

Here it is assumed that x is independent of y and z. Since the missing data are multinomial, the complete data pdf can be expressed as in (14.11), where v_{s_x}, w_{s_y}, and $z_{s'_y}$ are the components of v, x, and z associated with the Gaussians s_x, s_y, and s'_y, respectively:

$$\begin{aligned} p(x, y, v, w, z) \approx & \prod_{s_x} \prod_{s_y} \left[p(v_{s_x} = 1, w_{s_y} = 1) p(x|v_{s_x} = 1, w_{s_y} = 1) \right]^{v_{s_x} w_{s_y}} \\ & \times \prod_{s_x} \prod_{s_y} \prod_{s'_y} \left[p\left(v_{s_x} = 1, w_{s_y} = 1, z_{s'_y} = 1\right) \right. \\ & \left. p\left(y_t \middle| v_{s_x} = 1, w_{s_y} = 1, z_{s'_y} = 1\right)\right]^{v_{s_x} w_{s_y} z_{s'_y}}. \end{aligned} \tag{14.11}$$

Once the complete data pdf is obtained, the EM algorithm is applied iteratively in two steps: the expectation (E) step, which estimates the expected values of the missing data, and the maximization (M) step, which obtains the parameters of the cross-probability GMM using the estimated missing data.

14.4.1 The E Step

To evaluate the E step, the function $Q(\Theta|\Theta^{(k)})$ is defined as $Q(\Theta|\Theta^{(k)}) = E\left[\log(p(X, Y, V, W, Z|\Theta))\middle|X, Y, \Theta^{(k)}\right]$, where the operator $E[\,]$ is the expected value, k is the iteration index, and Θ includes all the unknown parameters of the cross-probability GMM we pretend to estimate. So, $Q(\Theta|\Theta^{(k)})$ is expressed as

$$\begin{aligned} Q\left(\Theta|\Theta^{(k)}\right) = & \sum_n \sum_{s_x} \sum_{s_y} \left(v_{s_x} w_{s_y}\right)^{(k)} \left[\log(p(s_x)p(s_y)) + \log\left(p\left(x_n \middle| v_{s_x} = 1, w_{s_y} = 1\right)\right)\right] \\ & + \sum_n \sum_{s_x} \sum_{s_y} \sum_{s'_y} \left(v_{s_x} w_{s_y} z_{s'_y}\right)^{(k)} \left[\log\left(p(s_x)p(s_y)p\left(s'_y|s_x, s_y\right)\right)\right. \\ & \left. + \log\left(p\left(y_n \middle| v_{s_x} = 1, w_{s_y} = 1, z_{s'_y} = 1\right)\right)\right], \end{aligned} \tag{14.12}$$

$$\left(v_{s_x} w_{s_y}\right)^{(k)} \approx E[v_{s_x}|x_n] E\left[w_{s_y}|y_n\right], \tag{14.13}$$

$$\left(v_{s_x} w_{s_y} z_{s'_y}\right)^{(k)} \approx \left(v_{s_x} w_{s_y}\right)^{(k)} E\left[z_{s'_y}|y_n, v_{s_x}, w_{s_y}, \Theta^{(k)}\right], \tag{14.14}$$

$$E\left[z_{s'_y}|y_n,v_{s_x},w_{s_y},\Theta^{(k)}\right] = \frac{p\left(s'_y|s_x,s_y\right)^{(k)} N\left(y_n\middle|\mu^{(k)}_{s_x,s_y,s'_y},\Sigma^{(k)}_{s_x,s_y,s'_y}\right)}{\sum\limits_{s'_y} p\left(s'_y|s_x,s_y\right)^{(k)} N\left(y_n\middle|\mu^{(k)}_{s_x,s_y,s'_y},\Sigma^{(k)}_{s_x,s_y,s'_y}\right)}, \tag{14.15}$$

where it is assumed that the variables v_{s_x} and w_{s_y} are independent, $E\left[v_{s_x}|x_n,y_n,\Theta^{(k)}\right] \approx E\left[v_{s_x}|x_n\right]$ and $E\left[w_{s_y}|x_n,y_n,\Theta^{(k)}\right] \approx E\left[w_{s_y}|y_n\right]$. $E\left[z_{s'_y}|y_n,\right.$ and $v_{s_x},w_{s_y},\Theta^{(k)}.]$ is estimated with (14.8) and (14.9) as (14.15), and $E[v_{s_x}|x_n]$ and $E\left[w_{s_y}|y_n\right]$ are computed in a similar way with (14.1) and (14.2), and with (14.3) and (14.4), respectively, assuming that there is only one basic environment. Although, in this work, to simplify, $E[v_{s_x}|x_n]$ and $E\left[w_{s_y}|y_n\right]$ values are 1, if the corresponding Gaussians are the most probable ones, and 0 in any other case (hard Gaussian estimation approach).

14.4.2 The M Step

To obtain the maximum likelihood estimates for the unknown parameters of the cross-probability GMM, $Q\left(\Theta|\Theta^{(k)}\right)$ is maximized with respect to them. So, the corresponding expressions for the $(k+1)$th iteration are

$$p\left(s'_y|s_x,s_y\right)^{(k+1)} = \frac{\sum\limits_n \left(v_{s_x}w_{s_y}z_{s'_y}\right)^{(k)}}{\sum\limits_{s'_y}\sum\limits_n \left(v_{s_x}w_{s_y}z_{s'_y}\right)^{(k)}}, \tag{14.16}$$

$$\mu^{(k+1)}_{s_x,s_y,s'_y} = \frac{\sum\limits_n \left(v_{s_x}w_{s_y}z_{s'_y}\right)^{(k)} y_n}{\sum\limits_n \left(v_{s_x}w_{s_y}z_{s'_y}\right)^{(k)}}, \tag{14.17}$$

$$\Sigma^{(k+1)}_{s_x,s_y,s'_y} = \frac{\sum\limits_n \left(v_{s_x}w_{s_y}z_{s'_y}\right)^{(k)}\left(y_n-\mu^{(k)}_{s_x,s_y,s'_y}\right)\left(y_n-\mu^{(k)}_{s_x,s_y,s'_y}\right)^{\mathrm{T}}}{\sum\limits_n \left(v_{s_x}w_{s_y}z_{s'_y}\right)^{(k)}}. \tag{14.18}$$

As it has been indicated, for MEMLIN, the cross-probability GMM parameters have to be estimated independently for each basic environment using the labelled training corpus $\left(X_{Tr,e},Y_{Tr,e}\right)$. So, the expressions (14.8) and (14.9) are transformed into

$$p\left(y_t\middle|s_x,s^e_y,e\right) = \sum_{s'_y=1}^{C''} p\left(y_t\middle|s_x,s^e_y,s'_y,e\right)p\left(s'_y\middle|s_x,s^e_y,e\right), \tag{14.19}$$

$$p\left(y_t\middle|s_x, s_y^e, s'_y, e\right) = N\left(y_t; \mu_{s_x,s_y^e,s'_y}, \Sigma_{s_x,s_y^e,s'_y}\right). \tag{14.20}$$

Here μ_{s_x,s_y^e,s'_y}, Σ_{s_x,s_y^e,s'_y}, and $p\left(s'_y\middle|s_x, s_y^e, e\right)$ are the mean vector, the diagonal covariance matrix, and the a priori probability associated with s'_y Gaussian of the cross-probability GMM associated with s_x and s_y^e. So, $p\left(s_x\middle|y_t, e, s_y^e\right)$ can be obtained as

$$p\left(s_x\middle|y_t, e, s_y^e\right) = \frac{p\left(y_t\middle|s_x, s_y^e, e\right)}{\sum_{s_x} p\left(y_t\middle|s_x, s_y^e, e\right)}. \tag{14.21}$$

14.5 Normalized Space Acoustic Models

Feature vector normalization techniques try to map the noisy feature vectors to the clean space. However, this mapping is not perfect and a new normalized space is created, which is different from the clean one. Thus, a further improvement can be obtained adapting the clean acoustic models toward the normalized space. For this purpose, the noisy training data are normalized in the same way as testing data and the original clean acoustic models are adapted with those data toward the new normalized space. If there are enough data, maximum likelihood (ML) algorithm can be used, but a model adaptation method should be applied otherwise (maximum *a posteriori,* MAP [5], MLLR [7], etc.). In this work, once the MEMLIN normalized space acoustic models are obtained, the normalized testing data can be recognized directly with them.

14.6 Discussion of Results

14.6.1 Results with SpeechDat Car Database

To observe the performance of the cross-probability GMM proposed in a real, dynamic, and complex environment, a set of experiments were carried out using the Spanish SpeechDat Car database [9]. Seven basic environments were defined: car stopped, motor running (E1), town traffic, windows close and climatizer off (silent conditions) (E2), town traffic and noisy conditions, windows open and/or climatizer on (E3), low speed, rough road, and silent conditions (E4), low speed, rough road, and noisy conditions (E5), high speed, good road, and silent conditions (E6), and high speed, good road, and noisy conditions (E7).

The clean signals are recorded with a CLose talK (CLK) microphone (Shure SM-10A), and the noisy ones are recorded by a hands-free (HF) microphone placed on the ceiling in front of the driver (Peiker ME15/V520-1). The SNR range for CLK signals goes from 20 to 30 dB, and for HF ones goes from 5 to 20 dB. For speech recognition, the feature vectors are composed of the 12 MFCCs, the energy, first and second derivatives, giving a final feature vector of 39 coefficients computed every 10 ms using a 25-ms Hamming window. On the other hand, in this work, the feature vector normalization methods are applied only to the 12 MFCCs and energy, whereas the derivatives are computed over the normalized static coefficients.

The recognition task is isolated and continuous digits recognition. The acoustic models are composed by 16-state 3 Gaussian continuous density HMM to model the 10 Spanish digits and 2 silence models for long (three-state 6 Gaussian continuous density HMM) and interword (one-state 6 Gaussian continuous density HMM) silences are used.

The word error rate (WER) baseline results for each basic environment are presented in Table 14.1, where MWER is the mean WER computed proportionally to the number of words in each basic environment. Cepstral mean normalization is applied to testing and training data. "Train" column refers to the signals used to obtain the corresponding acoustic HMMs: CLK if they are trained with all clean training utterances and HF if they are trained with all noisy ones. HF indicates that specific acoustic HMMs for each basic environment are applied in the recognition task (environment match condition). "Test" column indicates which signals are used for recognition: clean, CLK, or noisy, HF.

Table 14.1 shows the effect of real car conditions, which increases the WER in all the basic environments ("Train" CLK, "Test" HF), concerning the rates for clean conditions ("Train" CLK, "Test" CLK). When acoustic models are retrained using all basic environment signals ("Train" HF) MWER decreases. Finally, and in spite of the high WER reached for the basic environment E7 due to the reduced number of training utterances, 3.42% of MWER is obtained for environment match condition.

Figure 14.3 shows the mean improvement in WER (MIMP) in % for MEMLIN and MEMLIN with cross-probability model based on GMM

Table 14.1 WER baseline results, in percentage (%), from the different basic environments (E1, . . ., E7)

Train	Test	E1	E2	E3	E4	E5	E6	E7	MWER (%)
CLK	CLK	0.95	2.32	0.70	0.25	0.25	0.32	0.00	0.91
CLK	HF	3.05	13.29	15.52	27.32	31.36	35.56	53.06	21.49
HF	HF	3.81	6.86	3.50	3.76	4.96	4.44	3.06	4.63
HF	HF	1.14	4.37	1.68	2.13	2.10	2.06	23.13	3.42

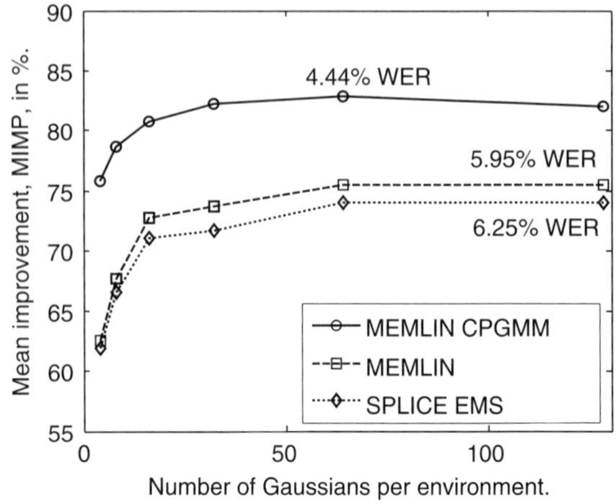

Fig. 14.3 Mean improvement in WER, MIMP, in % for MEMLIN, MEMLIN with cross-probability model based on GMM, MEMLIN CPGMM, and SPLICE with environmental model selection, SPLICE EMS

(MEMLIN CPGMM). Also the results with SPLICE with environmental model selection (SPLICE EMS) [4] are included. MIMP is computed as

$$\mathrm{MIMP} = \frac{100(\mathrm{MWER} - \mathrm{MWER}_{\mathrm{CLK-HF}})}{\mathrm{MWER}_{\mathrm{CLK-CLK}} - \mathrm{MWER}_{\mathrm{CLK-HF}}}, \tag{14.22}$$

where $\mathrm{MWER}_{\mathrm{CLK\text{-}CLK}}$ is the mean WER obtained with clean conditions (0.91 in this case), and $\mathrm{MWER}_{\mathrm{CLK\text{-}HF}}$ is the baseline value of 21.49. So, a 100% MIMP would be achieved when MWER equals the one obtained under clean conditions. The cross-probability GMMs are composed by two Gaussians for each pair of clean and noisy Gaussians. The important improvement of MEMLIN CPGMM concerning MEMLIN can be observed: from 62.57 to 75.79% with four Gaussians per basic environment and from 74.08 to 82.86% with 64 Gaussians.

Although the number of Gaussians to model the basic environments could be the same for MEMLIN and MEMLIN CPGMM, the computing time is not the same. To reduce it, only the cross-probability GMMs of the most probable pairs of Gaussians could be computed in normalization. Some experiments were carried out considering this alternative, showing that similar results can be obtained computing only a reduced number of pair of Gaussians [2].

Table 14.2 Best MWER and MIMP obtained with MEMLIN and MEMLIN CPGMM and matched acoustic conditions

Train	Test	MWER (%)	MIMP (%)
HF MEMLIN 64	HF MEMLIN 64	1.67	96.33
HF MEMLIN CPGMM 128	HF MEMLIN CPGMM 128	1.47	97.27

Table 14.2 shows the corresponding matching condition results (MWER and MIMP) when normalized acoustic models are used (clean and noisy condition results, Train CLK, Test CLK and Train HF, Test HF, can be observed in Table 14.1 to compare). In Train HF MEMLIN and Train HF MEMLIN CPGMM, the noisy training data normalized with MEMLIN or MEMLIN CPGMM are used to retrain the corresponding new acoustic models with the ML algorithm. The number of Gaussians per basic environment is included next to the normalization techniques and for MEMLIN CPGMM, the noisy feature vectors for each pair of Gaussian as s_x and s_y^e are modelled with two components (there is no significant differences in recognition if the basic environments are modelled with different number of Gaussians). Clearly there are significant improvements when normalized space acoustic models are used. It can be observed that the improvement with respect to using clean acoustic models is significant (4.44 and 5.95% of MWER for MEMLIN CPGMM and MEMLIN, respectively), and the comparison is even satisfactory if we compare the results with the ones reached with environment match condition ("Train" HF, "Test" HF and "Train" HF, "Test" HF). This is because the normalized space is not as heterogeneous as the noisy one and the training process can be more effective.

14.6.2 Results with Aurora2 Database

Aurora2 database [6] is built from TIDigits database utterances that have been digitally corrupted by passing them through a linear filter and/or by adding different types of noises at SNRs ranging from 20 to −5 dB. This does not define a real environment because not all kind of degradations are included, i.e., Lombard effect [8]; but, in spite of this weakness, Aurora2 is one of the most used database and it is almost a standard database to compare different techniques.

In this work, the MEMLIN and MEMLIN CPGMM parameters were trained using identical utterances from the clean training set and the multi-condition training set. This tunes the normalization parameters on the noise types from set A, keeping sets B and C as unseen conditions. Although the results for the three sets were obtained, in this work we only present the results with car noise-contaminated signals, which is considered as testing corpus and it is marked as HF to maintain the nomenclature. The parameters for speech

Table 14.3 Best results obtained with MEMLIN (M) and MEMLIN CPGMM (M C) with car noise-contaminated signals of Aurora2 database

Train	Test	−5 dB	0 dB	5 dB	10 dB	15 dB	20 dB	Clean	MWER (%)	MIMP (%)
CLK	HF	6.83	10.71	30.75	63.53	88.55	97.08	99.05	58.12	–
CLK	HF M 64	24.58	50.76	78.68	92.53	97.26	98.33	99.25	83.51	62.49
CLK	HF M C 64	26.67	55.53	82.98	94.40	97.53	98.51	99.25	85.79	67.52

recognition (acoustic models and feature vectors) are obtained in the same way as it is indicated in Section 14.6.2.

The recognition results obtained with Aurora2 database are presented in Table 14.3. It can be observed that MEMLIN (M) and MEMLIN CPGMM (M C) maintain the satisfactory performance, obtaining a mean improvement of 62.49 and 67.52%, respectively (the improvement is computed in this case as ETSI recommendation).

14.7 Conclusions

In this chapter, we have focussed on an approach of MEMLIN where the cross-probability model is estimated by modelling the noisy feature vectors associated with each pair of Gaussians from the clean and the noisy basic environment spaces with a GMM. MEMLIN obtains an improvement in WER of 75.53% with 128 Gaussians per environment with SpeechDat Car database in Spanish, whereas MEMLIN with cross-probability model based on GMM reaches 82.86% for 64 Gaussians to model each basic environment. If we consider Aurora2 database, and the recognition test is composed only by the car noise-corrupted signals, the improvements are, modelling each basic environments with 64 Gaussians, 62.49 and 67.52%, respectively. On the other hand, in order to reduce the mismatch between normalized feature vectors and clean acoustic models, we propose to obtain acoustic models which represent the normalized space. Applying this procedure to SpeechDat Car database, important improvements are obtained: 96.33 and 97.27% if the normalization technique is MEMLIN or MEMLIN CPGMM with 64 and 128 Gaussians per basic environment, respectively.

Acknowledgments This work has been supported by the Spanish National Project TIN 2005-08660-C04-01.

References

1. L. Buera, E. Lleida, A. Miguel and A. Ortega, "Multi-environment models based linear normalization for robust speech recognition in car environments", Proceedings of the ICASSP, vol. 1, pp. 1013–1016, 2004.
2. L. Buera, E. Lleida, J. A. Nolazco-Flores, A. Miguel and A. Ortega, "Time-dependent cross-probability model for multi-environment model based linear normalization", Proceedings of the ICSLP, 2006.
3. A.P. Dempster and D.B. Rubin, "Maximum likelihood from incomplete data via EM algorithm", Journal of the Royal Statistical Society, vol. 9, no. 1, pp. 1–37, 1977.
4. J. Droppo, L. Deng and A. Acero, "Evaluation of the SPLICE algorithm on the Aurora 2 database", Proceedings of Eurospeech, pp. 217–220, 2001.
5. J.L. Gauvain and C.-H. Lee, "Maximum a posteriori estimation for multivariate Gaussian mixture observations of Markov chains", IEEE Transactions SAP, vol. 2, pp. 291–298, 1994.

6. H.G. Hirsch and D. Pearce, "The aurora experimental framework for the performance evaluations of speech recognition systems under noisy conditions", Proceedings of the ISCA ITRW ASR2000, 2000.
7. C.J. Leggetter and P.C. Woodland, "Maximum likelihood linear regression for speaker adaptation of continuous density hidden Markov models", Computer Speech and Language, vol. 9, no. 2, pp. 171–185. ISSN 0885-2308.
8. E. Lombard, "Le signe de lélevation de la voix", Ann. Maladies Oreille, Larynx, Nez, Pharynx, vol. 37, pp. 101–119, 1911.
9. A. Moreno, B. Lindberg, C. Draxler, G. Richard, K. Choukri, S. Euler and J. Allen, "Speechdat-car, a large speech database for automotive environments", Proceedings of LREC, vol. 2, pp. 895–900, 2000.
10. L. Neumeyer and M. Weintraub, "Robust speech recognition in noise using adaptation and mapping techniques", Proceedings of ICASSP, vol. 1, pp. 141–144, 1995.
11. A. Sankar and C. Lee, "A maximum-likelihood approach to stochastic matching for robust speech recognition", IEEE Transactions SAP, vol. 4, pp. 190–202, 1996.
12. R. Stern, B. Raj and P. J. Moreno, "Compensation for environmental degradation in automatic speech recognition", ESCA Tutorial and Research Workshop on Robust Speech Recognition for Unknown Communication Channels, pp. 33–42, 1997.

15
Robust Feature Combination for Speech Recognition Using Linear Microphone Array in a Car

Yasunari Obuchi and Nobuo Hataoka

Abstract When speech recognition is performed in a car environment, there are two important robustness issues that should be taken into account. The first robustness is related to the noisy acoustic condition, and it has been one of the most popular research topics of in-vehicle speech recognition. In contrast, the second robustness, which is related to unstable calibration of the audio input, has not attracted much attention. Consequently, the performance of speech recognition would degrade greatly in a real application if the input device such as a microphone array is badly calibrated. We propose robust feature combination in the MFCC domain using speech inputs from a linear microphone array. It realizes robust (from both the noise and the calibration viewpoints) and practical speech recognition applications in car environments. Even a simple MFCC averaging approach is effective, and a new algorithm, hypothesis-based feature combination (HBFC), improves the performance. We also extend cepstral variance normalization as variance re-scaling, which makes the feature combination approach more robust. The advantages of the proposed algorithms are confirmed by the experiments using the data recorded in a moving car.

Keywords Speech recognition · Microphone array · MFCC · Feature combination · Hypothesis · Variance normalization · GMM

15.1 Introduction

There have been a lot of studies to realize robust speech recognition in noisy environments such as in cars and in public spaces (stations, stores, airports, etc.). Speech recognition using a microphone array is one of the successful approaches to realize such robustness. In most cases, microphone array techniques are implemented in the time domain or in the spectral domain to enhance the input signal to obtain better recognition performance. It is because those

Y. Obuchi (✉)
Central Research Laboratory, Hitachi Ltd., Japan

K. Takeda et al. (eds.), *In-Vehicle Corpus and Signal Processing for Driver Behavior*, DOI 10.1007/978-0-387-79582-9_15,

techniques are mainly focusing on the phase difference between the target signal and interfering noises. If the noise is directional, the phase difference can be measured clearly, and those typical microphone array approaches would work effectively. However, they are less effective for non-directional noises. In car environments, the speech recognition system is surrounded by various noise sources, and the directional noise assumption does not hold. The array processing algorithm should treat non-directional noises effectively.

The second problem is the robustness in the cases when calibration of the microphones and audio systems are not maintained well. When we use a microphone array in real applications, it is hard to maintain the stability of microphone characteristics, and it is the reason why there is a large discrepancy between the performance in the laboratory and in the real field. Hence the assumption that the power spectra of multiple inputs are identical does not hold, and we have various cepstral (MFCC) features corresponding to the multiple microphones. It is then reasonable to expect that combining them in the cepstral domain may improve the speech recognition performance. In fact, the Gaussian statistical nature of the cepstral features of speech suggests the isotropic nature of the cepstrum space and approves the effectiveness of feature combination in the cepstral domain. Moreover, since the speech can be modeled precisely in the cepstral domain using hidden Markov models (HMMs), we can take advantage of the prior knowledge about speech if we work in the cepstral domain.

In [1], we studied MFCC combination of the dual-microphone system and proposed hypothesis-based feature combination (HBFC). HBFC is one of the realizations to use the prior knowledge about speech, in which the input features and statistical model of speech are combined in the cepstral domain. In [2], the concept of feature combination in the cepstral domain was extended to a linear microphone array system and a problem of cepstral variance normalization was raised. In this chapter, these issues are investigated in detail and an approach to solve the cepstral variance normalization problem will also be presented.

15.2 MFCC Average and Variance Re-scaling

In [1], we showed that we can improve the speech recognition accuracy simply by averaging two MFCC sequences of the dual-microphone systems. Naturally, it can be extended to a multiple-input system as

$$\mathbf{x}_{\text{ave}} = \frac{1}{N}\sum_{i=0}^{N} \mathbf{y}i, \tag{15.1}$$

where $\mathbf{y}_{\mathrm{i}} = \{y_{itd} \mid 1 \leq t \leq T,\ 1 \leq d \leq D\}$ is the MFCC feature vector made from the observed signal by the ith microphone and $\mathbf{x}_{\text{ave}}$ is the corresponding combined feature vector. N is the number of microphones, T is the number of time frames, and D is the dimension of MFCC used.

However, our preliminary experiments had revealed that this simple averaging does not work well, especially in the case of large N. Furthermore, in the isolated word recognition experiments, it was found that the majority of the recognition errors converged into one specific word. It should be noted that this specific word coincides with the recognition output for the MFCC feature vector stream whose elements are all zero (assuming that the system has no rejection function). Taking into account that the arithmetic mean in the MFCC domain is almost equivalent to the geometric mean in the power spectral domain, MFCC average tends to take a smaller value in general. In particular, such a change occurs if the observed MFCC values have largely different values. Figure 15.1 shows the results of our preliminary experiments. Here, we first prepared two separate feature vector streams, one is identical to the single-input feature vectors and the other is the average of feature vectors obtained using seven microphones. We then calculated the ratio of the absolute values of two feature vectors and made a histogram. It is shown that the feature vector became smaller by averaging in more than 65% frames. The mean of the ratio was 0.96.

A simple solution to this problem is to multiply a fixed normalization factor to all the MFCC values

$$\mathbf{Z}_{\mathrm{ave}} = \alpha.\mathbf{X}_{\mathrm{ave}}. \tag{15.2}$$

If we use cepstral mean normalization (CMN) [3], the cepstral mean does not change by Eq. (15.2), and Eq. (15.2) can be interpreted as re-scaling of the cepstral variance. We already know that variance normalization improves the

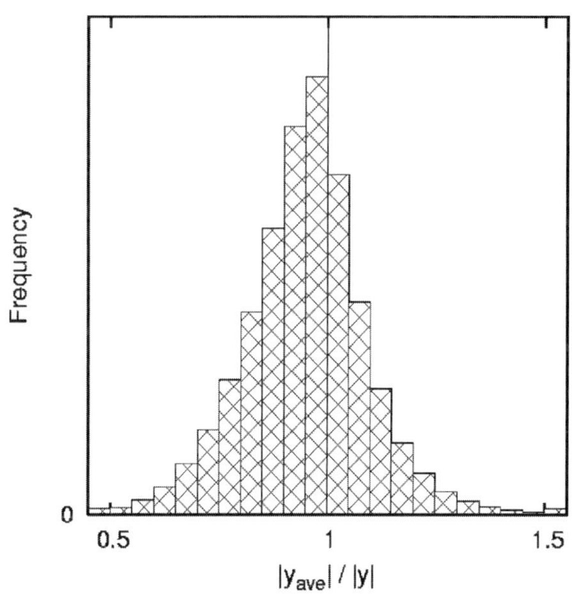

Fig. 15.1 Comparison of single input feature and averaged feature

speech recognition accuracy in noisy environment. Usually, inverse of the utterance-level standard deviation of each cepstral dimension is used as the normalization factor. However, it is not realistic to use the utterance-level statistics in real-time systems, and various approximation methods are introduced. In the framework described in this chapter, variance normalization described above is extended to deal with multiple-input channels.

15.3 GMM-Based Variance Normalization

As shown in Fig. 15.1, the cepstral variance of the averaged feature vector tends to be smaller than the original one. It can be compensated by Eq. (15.2), but the optimal value of the scaling factor α is not obvious from Eq. (15.2) itself. One of the solutions to this problem is to use prior knowledge about speech. In the proposed approach of automatic scaling factor estimation, we use Gaussian mixture models (GMMs) to estimate the effectiveness of a specific value of α. The proposed approach is referred to as GMM-based variance normalization (GVN).

It is natural to expect that the GMM score of the original feature vector is higher than that of the corrupted feature vector. We carried out a set of preliminary experiments, in which we added all frame-level GMM scores to obtain an utterance-level GMM score, using various values of α. Contrary to our expectation, it was revealed that the GMM score takes the maximum with a very small value of α such as 0.2, whether the original or the averaged feature vector was re-scaled. Experimental results indicate that either too large or too small GMM score means highly corrupted feature vectors. Therefore, our criteria must not be the absolute value of the GMM score, but the relative value of the GMM score compared with the single-input feature vector.

Equation (15.3) is the definition of GMM-based variance normalization, proposed in this chapter as the conclusion of the above discussion:

$$\alpha_{\text{opt}} = \arg\max_i(|S(\alpha_i \mathbf{x}_{\text{ave}}) - S(\mathbf{y}_0)|). \tag{15.3}$$

Here, the optimal value of α is chosen from a finite set of candidate values $\{\alpha_i\}$. $S(\mathbf{x})$ is the utterance-level GMM score of the feature vector sequence $\mathbf{x}$.

15.4 Hypothesis-Based Feature Combination of Multiple Inputs

As the successful applications of speech recognition research indicate, the speech signal can be modeled fairly accurately in the cepstral (MFCC) domain. Working in the MFCC domain has an advantage that we can use the prior knowledge about the speech model in a framework of the feature combination. From this viewpoint, we formerly proposed a feature combination method in the MFCC domain, referred to as hypothesis-based feature combination (HBFC) [1]. Figure 15.2 shows the schematic diagram of HBFC applied to more than two microphone inputs [2]. On the left-hand side, one channel (typically the central microphone) was chosen to be used in the first decoding process, and the obtained speech recognition hypothesis is used to synthesize the feature vector using the forced-alignment result and HMM. Feature synthesis is a simple procedure in which the mean vectors of the HMM state sequence (result of forced alignment) are simply concatenated. On the right-hand side, feature vectors of all the other channels are averaged. Finally, the outputs of two separate processes are combined by taking a simple weighted average:

$$\begin{aligned} \mathbf{x}_{\mathrm{HBFC}} &= w\mathbf{x}_{\mathrm{syn}} + (1-w)\mathbf{x}_{\mathrm{ave}} \\ &= w\mathbf{x}_{\mathrm{syn}} + \frac{1-w}{N-1}\sum_{k=1}^{N-1} \mathbf{y}k, \end{aligned} \tag{15.4}$$

where $\mathbf{x}_{\mathrm{syn}}$ is the output of the left-hand side of Fig. 15.2, $\mathbf{x}_{\mathrm{ave}}$ is the output of the right hand side of Fig. 15.2, N is the number of inputs, and w is the weight parameter. In our experiments, $N=7$ and $w=0.1$ are used. As shown in [1], the value of w should be small enough, because the synthesized feature vector always matches perfectly to one of the HMM states and produces relatively high likelihood score.

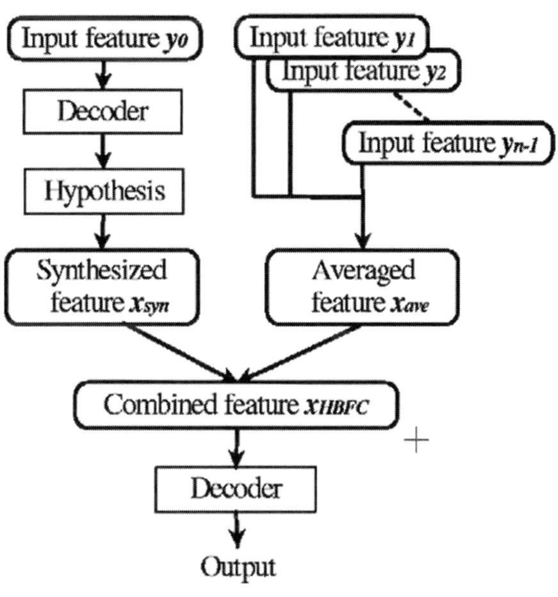

Fig. 15.2 Block diagram of multiple-input hypothesis-based feature combination

15.5 Experimental Results

15.5.1 Database and Setup

We have carried out several sets of experiments to evaluate various implementations of feature combination in the MFCC domain. The evaluation data were recorded in a real car which was running on urban roads in Tokyo. Therefore, the car stopped often due to the traffic jam, and the environmental noise includes road construction, sirens of the emergency vehicles, etc. The database also includes some data recorded on rainy days.

Our database is made of 3,620 utterances in total, uttered by 18 speakers (11 male and 7 female). The task is 152 Japanese POI (points of interest) isolated word recognition (IWR) to input the destination to the navigation system. The speaker sat in the passenger seat and was prompted each time to speak by a beep. The POI was selected spontaneously by the speaker each time. The speaker was asked to pick up a POI from the road atlas, and speak it after closing the atlas. The correct name of the registered POI is emphasized on the map, so that the out-of-vocabulary rate is kept small. However, there are some reading mistakes because Japanese "KANJI" characters have more than one pronunciation. Each utterance was roughly end-pointed by a fixed time-window from the beep position, so the utterance contains relatively high percentage of the non-speech segment.

The utterances were recorded using a microphone array mounted on the dashboard in front of the passenger seat. The microphone array is made of seven linearly located microphones. These microphones were numbered from 1 to 7 in the direction from the driver's side to the window side (right to left in Japan) and placed at intervals of 10, 5, 5, 5, 5, and 10 cm. The average SNR of all recorded data was estimated as −3.4 dB, but most of the noise exists in lower frequency range, and the estimated SNR increased to 10.0 dB after applying a bandpass filter with a 400–5500 Hz pass band. More details of the database can be found in [4].

For the recognition experiments, we prepared our original decoder and acoustic models. The acoustic models were made of Japanese triphones (3 states/model, 6 Gaussian mixtures/state) and trained using 16 hours of clean speech (read sentences), uttered by 120 speakers. All speech data were sampled by 16 kHz, converted to a 13-dimensional MFCC feature vector (including 0th MFCC) every 10 ms, and either CMN or cepstral mean and variance normalization (MVN) [5] was applied.

15.5.2 MFCC Average

Figure 15.3 shows the recognition results obtained by the averaged and rescaled MFCC features. The recognition rates of single input (CMN: 87.43% and MVN: 86.32%) and a delay-and-sum beamformer [6] (CMN: 88.20%) are shown by the horizontal lines for comparison.

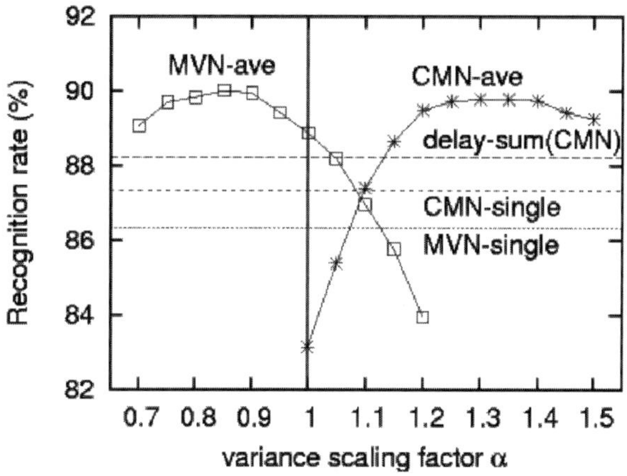

Fig. 15.3 Experimental results of MFCC average

When we look at the data points on the line of $\alpha = 1.0$, we can confirm the recognition rates without variance re-scaling, which were 83.15% (CMN) and 88.87% (MVN), respectively. Since the feature vectors after CMN have larger variety over microphones than MVN, the variance of their average tends to be smaller, and it results in the lower recognition rate. However, if we can use the optimal value of α (1.30 for CMN and 0.85 for MVN), the recognition rate would be much higher than the standard delay-and-sum beamformer.

Next, we have explored with GMM-based variance normalization. The results obtained with CMN are shown in Fig. 15.4. In this figure, "GMM score" means the sum of the utterance-level relative GMM scores:

$$S^{\rm rel}(\alpha) = \sum_{j=1}^{N} |S(\alpha \mathbf{x}_{{\rm ave},j}) - S(\mathbf{y}_{0,j})|, \tag{15.5}$$

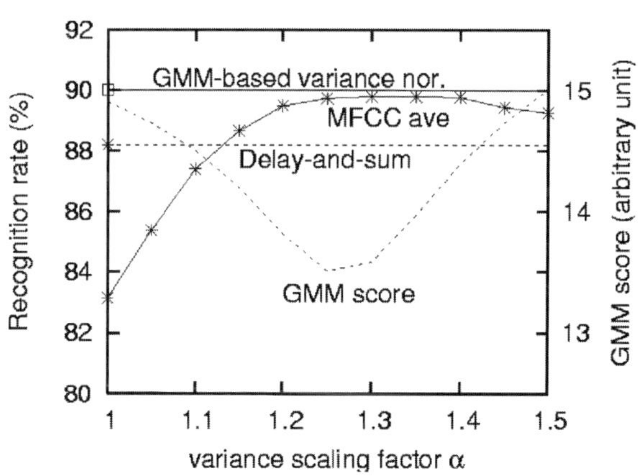

Fig. 15.4 Comparison of GMM scores, recognition rates for averaged features, and recognition rates of GMM-based variance normalization. All feature vectors were normalized by CMN

where subscript j represents each utterance and N is the number of test utterances. It is clear that the curves of the recognition rate of the average and the GMM score are almost symmetric. The GMM score takes the smallest value with $\alpha = 1.25$, with which the recognition rate is 89.72%. If we apply GMM-based variance normalization utterance by utterance, the recognition rate becomes 90.00%.

Figure 15.5 shows the equivalent results obtained with MVN. It is not as symmetric as in Fig. 15.4, but there is a similar tendency, and the recognition rate of the GMM-based variance normalization is 88.26%, which is still better than the delay-and-sum beamformer.

15.5.3 Hypothesis-Based Feature Combination

In the final set of experiments, we tried HBFC with variance re-scaling. Figure 15.6 shows the results obtained with HBFC and CMN. The symmetric nature of the GMM score and the recognition rate is preserved well. With a fixed scaling factor, the highest recognition rate of 89.93% was obtained with $\alpha = 1.5$, and the recognition rate of GMM-based variance normalization is still higher than it (90.00%).

Figure 15.7 shows the results obtained with HBFC and MVN. In this case, the lowest GMM score was obtained with α even lower than 0.7, and the two curves (GMM score and HBFC recognition rate) are not symmetric. Consequently, the recognition rate of GMM-based variance normalization is lower than the HBFC recognition rates near $\alpha = 1.0$. However, it is worth mentioning that the recognition rate of GMM-based variance normalization is higher than that of HBFC with $\alpha = 0.7$, where the GMM score is smallest (in this figure). The degradation of GMM-based variance normalization from the highest recognition rate of HBFC is 3.07 points.

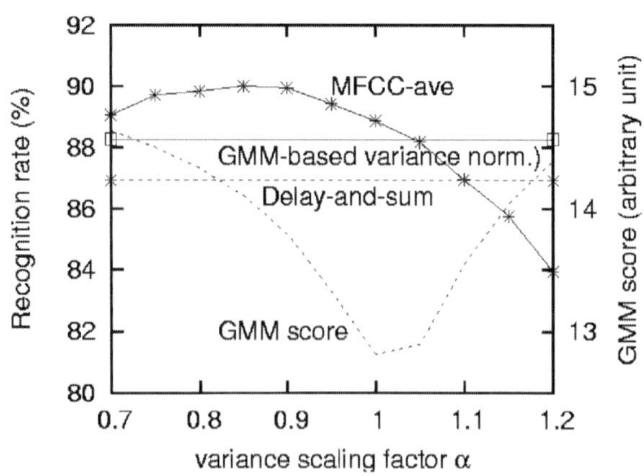

Fig. 15.5 Comparison of GMM scores, recognition rates for averaged features, and recognition rates of GMM-based variance normalization. All feature vectors were normalized by MVN

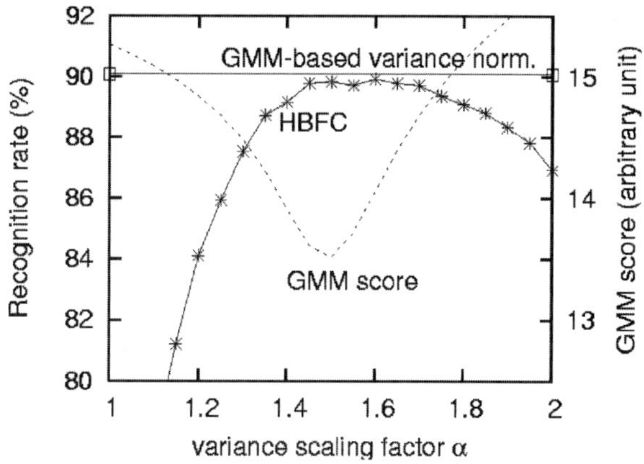

Fig. 15.6 Comparison of GMM scores, recognition rates for HBFC, and recognition rates of GMM-based variance normalization. All feature vectors were normalized by CMN

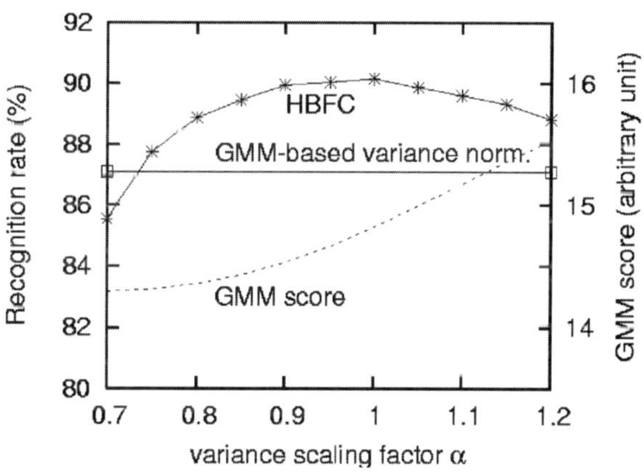

Fig. 15.7 Comparison of GMM scores, recognition rates for HBFC, and recognition rates of GMM-based variance normalization. All feature vectors were normalized by MVN

15.6 Conclusions

In this chapter, we have attempted to demonstrate that how speech recognition accuracy can be improved by various ways of feature combination in the MFCC domain. Simple averaging of MFCC features tends to lower the recognition rate, especially when only the cepstral means are normalized (CMN). However, such degradation can be explained by the fact that the averaged MFCC features tend to be smaller than the original MFCC feature, and it can be compensated by introducing variance re-scaling.

Next, we proposed a new algorithm to estimate the optimal value of the scaling factor, using the GMM score of the re-scaled feature vector and the single-input feature vector. Experimental results have shown that the proposed algorithm worked quite well if the feature vectors were normalized by CMN.

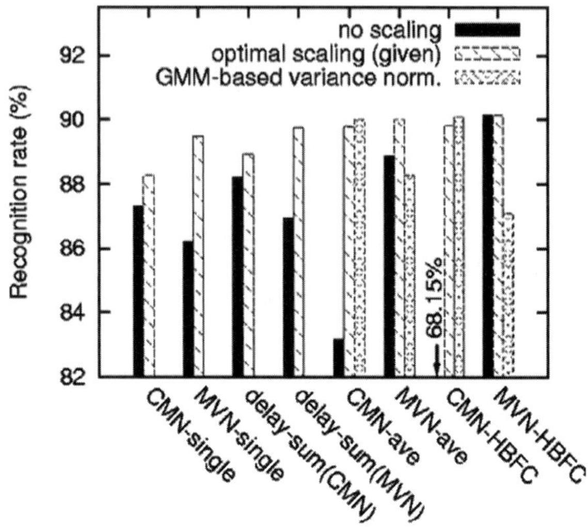

Fig. 15.8 Comparison of various algorithms evaluated in this chapter. The highest recognition rate in all is 90.14% with MVN-HBFC (optimal scaling and no scaling coincided)

Figure 15.8 shows the comparison of the algorithms evaluated in this chapter. As mentioned in [2], MVN-HBFC gives the highest recognition rate without variance re-scaling, but the problem with CMN-ave and CMN-HBFC were solved by the proposed algorithm.

Acknowledgments The authors are thankful to Professor Sadaoki Furui of Tokyo Institute of Technology and Professor Tetsunori Kobayashi of Waseda University for their valuable comments. This work was supported in part by the New Energy and Industrial Technology Development Organization (NEDO), Japan.

References

1. Y. Obuchi, "Hypothesis-based feature combination for dual-microphone speech recognition," Proc. HSCMA, Piscataway, NJ, USA, 2005.
2. Y. Obuchi and N. Hataoka, "Hypothesis-based feature combination of multiple speech inputs for robust speech recognition in automotive environments," Proc. Interspeech2006-ICSLP, Pittsburgh, PA, USA, 2006.
3. B. S. Atal, "Effectiveness of linear prediction characteristics of the speech wave for automatic speaker identification and verification," Journal of Acoustical Society of America, Vol. 55, No. 6, pp. 1304–1312, 1974.
4. Y. Obuchi and N. Hataoka, "Development and evaluation of speech database in automotive environments for practical speech recognition systems," Proc. Interspeech2006-ICSLP, Pittsburgh, PA, USA, 2006.
5. J. P. Openshaw and J. S. Mason, "On the limitations of cepstral features in noise," Proc. ICASSP, Adelaide, Australia, 1994.
6. W. Kellermann, "A self steering digital microphone array," Proc. ICASSP, Toronto, Canada, 1991.

16
Prediction of Driving Actions from Driving Signals

Toshihiko Itoh, Shinya Yamada, Kazumasa Yamamoto, and Kenji Araki

Abstract A spoken dialogue system for car-navigation systems may be able to provide more natural and smoother communications but it must also cause safety problems. One of these problems is distraction whereby machine operation and voice conversations influence the driver. Even the use of a simple speech interface may affect the driving operation. We consider that a spoken dialogue system which can understand the driver's situation and change its dialogue rhythm according to that situation would be safe as part of a car-navigation system. For this to be possible, the system needs to predict and recognize driver's actions from environmental information such as driving signals. In this chapter, we report the results of an experiment on predicting driver actions. The action prediction system uses HMM-based pattern recognition only on driving signals and does not use position information. Its best driving action prediction accuracy was 0.632.

Keywords Car-navigation systems · Distraction · Hidden Markov model · HMM-based pattern recognition · Dialogue rhythm · Driving action · Driving prediction · Driving recognition · Driving signal · Safety · Spoken dialogue · Spoken dialogue interface

16.1 Introduction

Conventional car-navigation systems usually have a remote controller and a touch panel as an input and use a display output. Recently, for safety reasons, speech interfaces have been watched with keen interest and car-navigation systems with speech command operation have appeared. We developed a spoken dialogue interface to make a car-navigation system easier to use. The spoken dialogue interface can respond with the dialogue rhythm of humans.

T. Itoh (✉)
Department of Information Science and Technology, Hokkaido University,
Sapporo, Japan
e-mail: t-itoh@media.eng.hokudai.ac.jp

K. Takeda et al. (eds.), *In-Vehicle Corpus and Signal Processing for Driver Behavior*, DOI 10.1007/978-0-387-79582-9_16,

Unfortunately, a spoken dialogue interface for car-navigation system may have problems whereby more natural and smoother communications actually adversely affect safety. One problem is distraction whereby machine operation and voice conversations influence driving operations [1, 2]. One of the biggest distractions is conversation on cellular phones. It has been pointed out that even the use of a simple speech interfaces may adversely affect driving operations [3].

In fact, we should beware of the situation in which the temporal restrictions enabling rhythmical communication in our system increase the danger of distraction. However, we also think that distractions exist when a dialogue partner does not understand or share the driver's situation. If a spoken dialogue system can understand the driver's situation and the dialogue rhythm of the system can be changed according to the driver's situation, a safe spoken dialogue interface can be created for a car-navigation system. For that to be possible, the system needs to predict and recognize the driver's actions from driving signals and other information.

Pentland et al. used dynamic Markov models to recognize human behaviors from sensory data and to predict human behaviors over the course of a few seconds [4]. Oliver et al. trained dynamical graphical models (HMMs and CHMMs) with experimental driving data to create models of seven different driver maneuvers: passing, changing lanes right and left, turning right and left, starting, and stopping [5].

Our research aims at the construction of a driver action prediction system that uses a simple and fast algorithm and simple data (signals) that can be acquired from unmodified cars, even though its prediction accuracy is low. (This aspect is different from other studies.)

In this chapter, we report the results of an experiment on driver action prediction. Our prediction system uses HMM-based pattern recognition only on driving signals and does not use position information.

16.2 Driving Signals

In this work, we have used driving signals in the CIAIR-HCC database [6] (the database of driver' commands to in-vehicle navigators).

16.2.1 Types of Driving Signals

Observable driving signals can be categorized into three groups [7]:

1. Driving behavioral signals (e.g., gas pedal pressure, brake pedal pressure, and steering angle)
2. Vehicle status signals (e.g., velocity, acceleration, and engine speed)
3. Vehicle position signals (e.g., following distance, relative lane position, and yaw angle).

Table 16.1 Signal data of driving behavior

Driving signal	Range
Gas pedal pressure	0–50 kgf / cm^2
Brake pedal pressure	0–50 kgf / cm^2
Steering angle	$-1800°$ to $1800°$
Engine speed	0–8000 rpm
Car velocity	0–120 km/h

Among these, we focused on driving behavioral signals and vehicle status signals that can be acquired from cars. We think that a vehicle position signals are very effective to predict driving actions, so the integration of such signals will be a future work.

16.2.2 Database

Driving behavioral signals were collected using a data collection vehicle (Toyota Regius) that was specially designed for data collection in the Center for Integrated Acoustic Information Research (CIAIR) project. Detailed information on this corpus can be found in [8]. Each driver drove the car on a city road, and 5-channel driving signals, 16-channel speech signals, 3-channel video signals, and global position data from GPS were recorded. The driving signals included force on gas and brake pedals, engine speed, car velocity, and steering angle (Table 16.1). These signals were sampled at 1 kHz, and each sample was expressed in 16 bits (the quantization level is 15 bits which use the data range from 0 to 32,767).

16.3 Predicting Driving Actions

16.3.1 Kinds of Driving Actions

As mentioned above, our goal is prediction of the driver's situation so that drivers can speak to navigation systems safely. For example, when a high-risk driving action is predicted to occur in a few seconds, the system would stop talking to the driver. Therefore, it is more important for the system to predict whether a high-risk driving action will occur in the near future than for it to recognize the current driving action.

We think that there are not many driving actions that should be predicted. These are as follows:

1. Straight ahead
2. Stop
3. Turn right

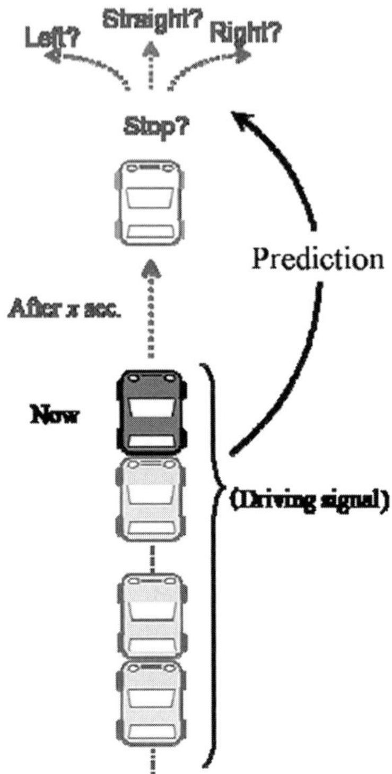

Fig.16.1 Driving action prediction

4. Turn left
5. Change lanes right
6. Change lanes left
7. Avoid obstacles

Figure 16.1 illustrates driving action prediction. Driving signals are continuously input to the system, and the system predicts what the driving action a few moments hence will be.

16.3.2 Methodology of Driving Action Prediction

We used an HMM-based statistical pattern recognition method for the prediction. HMM-based methods are widely used for recognition of time series like speech, and our driving action prediction can be viewed as recognition of time-series patterns appearing before the driving actions. To be accurate, the preparation action occurring before the actual driving action we want to predict is recognized, and the driving action is predicted.

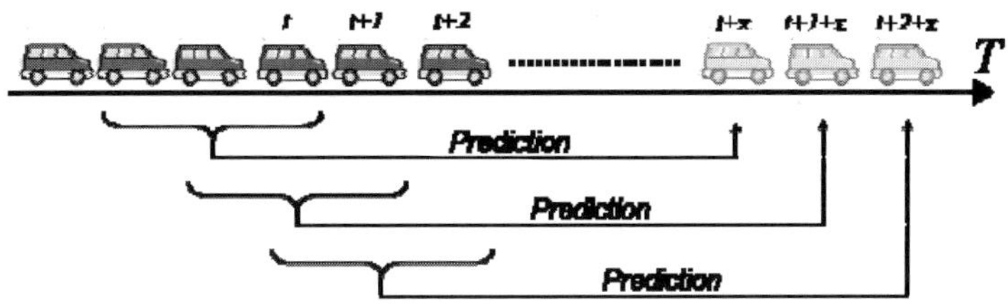

Fig.16.2 Our driving action prediction scheme

Therefore, we should model the preparation action preceding each driving action. The driving signals are time-series data, and HMMs have the ability to model such a time series. In speech recognition, an input is usually a sequence of n-dimensional vector data derived from a sequence of speech frames. For driving action prediction, we used 10-dimensional vector data including 5 driving signals (gas pedal pressure, brake pedal pressure, steering angle, engine speed, and car velocity) and 5 regressive coefficients.

The driving action prediction should be done frame by frame, as shown in Fig. 16.2. This is a kind of detection task called "word spotting" in the speech recognition field.

But as this is the first trial of the driving action prediction, we first had to confirm whether our strategy was effective or not. Thus, we manually extracted the periods preceding some driving actions and conducted an experiment to classify the intervals.

16.4 Experimental Data

16.4.1 Driving Action Labels

To predict the user's action described in Section 16.3.1, we manually detected the beginnings and ends of driving actions by viewing video and driving signals, and labeled the starts and ends of the individual actions by using an authoring tool.

Detailed features were also labeled for future research purposes. The kinds of the labels are shown in Table 16.2. "Stop" is labeled from the moment the car completely stops to when it begins to move again. In addition, labels were classified by the stopping purpose and stopping state. "Turn right", "Turn left", "Change lanes right", "Change lanes left", and "Avoid obstacles" durations begin when the steering wheel turns and end when the car begins to go straight again. For these actions, the labels must also be classified by purpose and state. In addition, we labeled "Passing" when the car avoided an obstacle (when a center of the car passed a center of the obstacle).

Table 16.2 Driving action labels

Action	Detailed	Label
Stop	Start straight	Ss
	End straight	Se
	Start turn	S[L,R]s
	End turn	S[L,R]e
	Start turning	S[L,R]Ds
	End turning	S[L,R]De
Turn right	Start	Rs
	End	Re
	Start after stop	RSs
	End after stop	RSe
Turn left	Start	Ls
	End	Le
	Start after stop	LSs
	End after stop	LSe
Change lanes right	Start	CRs
	End	CRe
Change lanes left	Start	CLs
	End	CLe
Avoid obstacles	Start	As
	Passing	A
	End	Ae

16.4.2 Training Data and Test Data

We made training and test data for the driving action prediction experiments by using the following process (Fig. 16.3). Our task was to predict the driving action occurring at time X, and the prediction should be done Y seconds before the occurrence. We thus assumed that the system can use the past Z-second signals to make its predictions.

We have extracted the driving signals from $(X–Y–Z)$ to $(X–Y)$. We also extracted regions without any special driving actions following the signals as the data from which the system should predict the “Straight ahead” action.

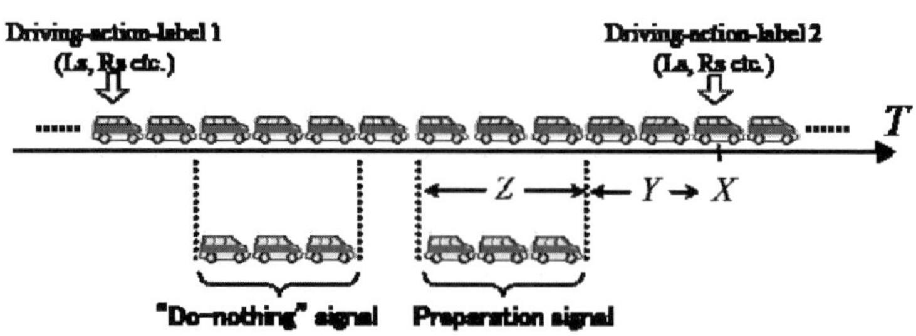

Fig.16.3 Extraction of experimental driving signals

Table 16.3 Extracted experimental driving action data

Driving action	Total data #
Straight ahead	1570
Stop	764
Turn right	66
Turn left	184
Change lanes	141
Avoid obstacles	1426

Similar to above, we have obtained "preparation signals" for driving actions from the driving data of 110 subjects. The preparation signals from 70 subjects were used as training data and those from the remaining 40 were used as test data. Table 16.3 lists the details of the data. This chapter reports only the results of a preliminary experiment in which the action happened just after the prediction (in other words, Y was set to 0).

16.5 Results of Prediction Experiments

The de facto standard HTK Toolkit [9] was used for training all the HMMs and the driving action prediction experiments.

16.5.1 Prediction Performance for Different Driving Signal Input Durations (Experiment 1)

First, we have investigated the change in prediction performance caused by differences in the input duration of driving signals. Table 16.4 shows the experimental conditions. We prepared 1, 5, and 10 seconds worth of driving signals for the prediction. Signals from 20 subjects were used as training data and those from 40 others were used as test data.

Table 16.5 shows the accuracy of the experiment. The prediction accuracy tended to improve as the input signal duration got longer. The performances with 5 seconds of input and with 10 seconds of input were almost the same. But

Table 16.4 Conditions of experiment 1

HMM	Ergodic HMM
Number of states	4
Number of mixtures	2, 4, 8, 16, 32
Feature parameters	5 driving signals + 5 Δ driving signals
Duration of input	1, 5, 10 seconds
# Training data	20 subjects
# Test data	Other 40 subjects

Table 16.5 Accuracy of experiment 1

No. of mixtures	Duration of input		
	1 second	5 seconds	10 seconds
2	0.470	0.502	0.522
4	0.488	0.554	0.521
8	0.546	0.587	0.544
16	0.559	0.554	0.579
32	0.550	0.557	0.527

when the input duration was 10 seconds, the prediction accuracy of the 32 mixtures decreased in comparison with the 16 mixtures.

We think that the training data in this experiment was not enough. Furthermore, when the duration of the input was 10 seconds, the training time of HMMs increased considerably. Hence, we think that the input signal duration should be set at about 5 seconds.

Table 16.6 shows the confusion matrix of the prediction result with the highest accuracy. "AO" means avoid obstacles, "CL" change lanes (left or right), "TL" turn left, "TR" turn right, and "SP" stop.

16.5.2 Prediction Performance for Different Amounts of Training Data (Experiment 2)

We have studied the change in prediction performance caused by varying the amount of training data.

Table 16.7 shows the experimental conditions, and Table 16.8 shows the accuracy of the experiment. The more training data were used, the higher accuracy we obtained. The highest accuracy was achieved by the model with 32 mixtures trained using 10-second duration data of all subjects.

Table 16.9 shows the confusion matrix of the prediction with the highest accuracy in this experiment.

Table 16.6 Confusion matrix of experiment 1 (16 mixtures and 10 seconds input duration)

	AO	CL	TL	SA	TR	SP	Total	Precision	*F-value*
AO	196	15	19	117	2	1	351	0.558	0.577
CL	25	12	2	31	0	0	70	0.171	0.198
TL	12	8	17	79	2	2	123	0.138	0.180
SA	76	12	16	298	8	2	422	0.706	0.592
TR	8	1	12	21	9	8	62	0.145	0.207
SP	11	3	0	39	4	246	286	0.801	0.840
Total	328	51	66	585	25	259	1314		
Recall	0.598	0.235	0.258	0.509	0.360	0.950		0.579	

Table 16.7 Conditions of experiment 2

HMM	Ergodic HMM
Number of states	4
Number of mixtures	2, 4, 8, 16, 32
Feature parameters	5 driving signals + 5 Δ driving signals
Duration of input	5, 10 seconds
# Training data	20, 40, 70 subjects
# Test data	Other 40 subjects

Table 16.8 Accuracy of experiment 2

	Input duration				
	5 seconds # Training data		10 seconds # Training data		
No of mixtures	20 subjects	70 subjects	20 subjects	40 subjects	70 subjects
2	0.502	0.536	0.522	0.550	0.567
4	0.554	0.480	0.521	0.522	0.559
8	0.587	0.501	0.544	0.557	0.549
16	0.554	0.517	0.579	0.590	0.573
32	0.557	0.487	0.527	0.550	0.632

Table 16.9 Confusion matrix of experiment 2
(32 mixtures, 10 seconds input duration, training data from 70 subjects)

	AO	CL	TL	SA	TR	SP	Total	Precision	F-value
AO	227	13	9	128	2	1	380	0.597	0.641
CL	35	30	4	53	0	0	122	0.246	0.347
TL	9	3	43	73	1	2	131	0.328	0.437
SA	31	2	5	267	3	2	310	0.861	0.597
TR	14	1	5	38	17	8	83	0.205	0.315
SP	12	2	0	26	2	246	288	0.854	0.899
Total	328	51	66	585	25	259	1314		
Recall	0.692	0.588	0.652	0.456	0.680	0.950		0.632	

16.5.3 Prediction Performance that Considers Individuality of Driving

So far, we used training data of persons not included in the test data. Note, however, that the drivers have their own styles, and this is why we conducted the experiment in which two-thirds of the signals for one driver were used as training data and the other one-third as test data.

Table 16.10 shows the experimental conditions, and Table 16.11 shows the accuracy of the experiment. Compared with the results of the models trained

Table 16.10 Conditions of experiment 3

HMM	Ergodic HMM
Number of states	4
Number of mixtures	2, 4, 8, 16, 32
Feature parameters	5 driving signals + 5 Δ driving signals
Duration of input	5 seconds
# Training data	2/3 of 110 subjects
# Test data	1/3 of 110 subjects

Table 16.11 Accuracy of experiment 3

No. of mixtures	Accuracy
2	0.522
4	0.515
8	0.505
16	0.543
32	0.529

using 5 seconds data of 70 subjects in Table 16.8, the accuracy improved even though almost the same amount of data was used. This reveals that it is important to consider the effect of driver individuality.

Table 16.12 shows the confusion matrix of the prediction result with the highest accuracy in this experiment.

16.5.4 *Prediction Performance Using Only One Signal out of five Driving Signals (Experiment 4)*

Next, we investigated the contribution level to the prediction of an individual driving signal. We conducted an experiment in which only one of the driving signals (gas pedal pressure, brake pedal pressure, engine speed, car velocity, or steering angle) was used.

Table 16.12 Confusion matrix of experiment 3 (16 mixtures)

	AO	CL	TL	SA	TR	SP	Total	Precision	*F*-value
AO	180	12	9	134	2	2	339	0.531	0.570
CL	25	17	8	45	0	0	95	0.179	0.238
TL	35	11	39	89	9	1	184	0.212	0.317
SA	39	7	1	168	2	7	224	0.750	0.449
TR	10	1	5	35	7	2	60	0.117	0.169
SP	4	0	0	53	3	245	305	0.803	0.872
Total	293	48	62	524	23	257	1207		
Recall	0.614	0.354	0.629	0.321	0.304	0.953		0.543	

Table 16.13 Conditions of experiment 4

HMM	Ergodic HMM
Number of states	4
Number of mixtures	2, 4, 8, 16, 32
Feature parameters	Individual driving signals
Duration of input	5 seconds
# Training data	40 subjects
# Test data	40 subjects

Table 16.13 shows the experimental conditions, and Table 16.14 shows the accuracies. The contribution level to the prediction greatly depended on the signal. For driving action prediction, we noticed that steering angle, car velocity, and engine speed are very useful pieces of information, but that gas pedal pressure and brake pedal pressure are not. Certainly, the latter two signals vary more largely according to the driving situation than to the driving action. These signals may negatively affect the prediction.

Table 16.15 shows the confusion matrix of the prediction result with the highest accuracy.

Table 16.14 Accuracy of experiment 4

	Driving signal				
No. of mixtures	Gas	Brake	Engine	Velocity	Steering
2	0.203	0.209	0.352	0.436	0.238
4	0.225	0.303	0.370	0.431	0.341
8	0.243	0.278	0.377	0.387	0.333
16	0.348	0.205	0.356	0.437	0.349
32	0.201	0.206	0.362	0.438	0.424

Table 16.15 Confusion matrix of experiment 4 (32 mixtures and input signal is car velocity)

	AO	CL	TL	SA	TR	SP	Total	Precision	*F*-value
AO	174	18	27	163	3	5	390	0.446	0.506
CL	29	20	3	52	0	11	115	0.174	0.227
TL	31	7	20	63	8	1	130	0.154	0.207
SA	62	14	10	249	7	153	495	0.503	0.481
TR	1	2	1	10	5	11	30	0.167	0.189
SP	1	0	2	3	0	76	82	0.927	0.448
Total	298	61	63	540	23	257	1242		
Recall	0.584	0.328	0.317	0.461	0.217	0.296		0.438	

16.5.5 Prediction Performance Using Three Useful Signals out of Five

The gas pedal pressure and brake pedal pressure are easy to change a lot. These signals may negatively affect the driving action prediction. So we used only steering angle, car velocity, and engine speed.

Table 16.16 shows the experimental conditions, and Table 16.17 shows the accuracies. Prediction performance was the highest so far under the condition using 5 seconds input signal duration. Compared with the results of the models trained using 5 seconds data, the accuracy improved even though the less amount of data was used.

This result shows that because the influences of individuality and situation cause the fluctuation of driving actions to be large, HMMs cannot model individual driving actions if they take too many input signals.

Table 16.18 shows the confusion matrix of the prediction result with the highest accuracy in this experiment.

16.5.6 Prediction Performance Using Detailed Classification of Driving Actions (Experiment 6)

Above results indicate that a lot of variations in the patterns were included in the kind of action. We divided each driving action into patterns according to the similarity of driving signals by using a clustering tool ("Straight ahead" was

Table 16.16 Conditions of experiment 5

HMM	Ergodic HMM
Number of states	4
Number of mixtures	2, 4, 8, 16, 32, 64
Feature parameters	3 driving signals (engine, velocity, steering)
Duration of input	5 seconds
# Training data	40 subjects
# Test data	Other 40 subjects

Table 16.17 Accuracy of experiment 5

No. of mixtures	Accuracy
2	0.475
4	0.525
8	0.537
16	0.571
32	0.584
64	0.594

Table 16.18 Confusion matrix of experiment 5 (64 mixtures)

	AO	CL	TL	SA	TR	SP	Total	Precision	*F*-value
AO	244	29	16	181	3	3	476	0.513	0.600
CL	9	7	3	22	0	0	41	0.171	0.151
TL	19	1	39	21	4	0	84	0.464	0.517
SA	60	13	9	264	11	26	383	0.689	0.545
TR	3	0	0	9	4	4	20	0.200	0.178
SP	2	2	0	88	3	233	328	0.710	0.785
Total	337	52	67	585	25	266	1332		
Recall	.724	.135	.582	.451	.160	.876		.594	

divided into four sub-classes, "Stop" was composed of three, "Turn right" was two, "Turn left" was four, "Change lanes right and left" was four, and "Avoid obstacles" was four.). An HMM of each sub-class was trained using the only driving signals included in the sub-class. The class whose sub-class had the highest likelihood was then selected as the predicted action.

Table 16.19 shows the experimental conditions and Table 16.20 shows the accuracies. Accuracy was the highest so far under the condition using the 5 seconds input signal duration.

Table 16.21 shows the confusion matrix of the prediction result with the highest accuracy in this experiment.

Table 16.19 Conditions of the experiment 6

HMM	Ergodic HMM
The number of state	4
The number of mixture	2, 4, 8, 16, 32, 64
Feature parameters	5 driving signals + 5 Δ driving signals
Duration of the input	5 seconds
# Training data	70 subjects
# Test data	Other 40 subjects

Table 16.20 Accuracy of experiment 6

No. of mixtures	Accuracy
2	0.487
4	0.533
8	0.561
16	0.593
32	0.615
64	0.612

Table 16.21 Confusion matrix of experiment 6 (32 mixtures)

	AO	CL	TL	SA	TR	SP	Total	Precision	*F*-value
AO	266	31	16	197	4	1	515	0.517	0.627
CL	13	6	1	19	0	0	39	0.154	0.133
TL	14	2	32	54	3	1	106	0.302	0.370
SA	31	8	13	255	9	12	328	0.777	0.559
TR	3	0	3	2	5	0	13	0.385	0.263
SP	6	4	2	58	4	251	335	0.772	0.851
Total	333	51	67	585	25	265	1326		
Recall	0.799	0.118	0.478	0.436	0.200	0.947		0.615	

16.6 Conclusion

We reported the results of an experiment on predicting driver's actions. The prediction system uses HMM-based pattern recognition only on driving signals and does not use position information. The best driving action prediction accuracy was 0.632. Considering that the prediction system used only a very simple framework and simple input signals, this result is very promising.

We believe that vehicle position signals will be very useful for predicting driving actions, and that information about the position signals can act as a grammatical constraint in speech recognition. We will try other methods to predict driving actions.

References

1. J. D. Lee, T. L. Brown, B. Caven, S. Haake, K. Schmidt, "Does a speech-based interface for an in-vehicle computer distract drivers?", Proc. World Congress on Intelligent Transport System, 2000.
2. http://www-nrd.nhtsa.dot.gov/departments/nrd-13/DriverDistraction.html.
3. http://www-nrd.nhtsa.dot.gov/departments/nrd-13/driver-distraction/Welcome.htm.
4. A. Pentland, A. Liu, "Modeling and prediction of human behavior", Neural Computation, vol. 11, pp. 229–242, 1999.
5. N. Oliver, A.P. Pentland, "Driver behavior recognition and prediction in a SmartCar", Proc. SPIE Aerosense 2000, Enhanced and Synthetic Vision, 2000.
6. http://db.ciair.coe.nagoya-u.ac.jp/eng/index.html
7. Y. Nishiwaki, K. Ozawa, T. Wakita, C. Miyajima, K. Itou, K. Takeda, "Driver identification based on spectral analysis of driving behavioral signals", Proc. Biennial on DSP for In-Vehicle and Mobile Systems, 2005.
8. N. Kawaguchi, S. Matsubara, K. Takeda, F. Itakura, "Multimedia data collection of in-car speech communication", Proc. EUROSPEECH 2001, pp. 2027–2030, 2001.
9. S. Young, D. Kershaw, J. Odell, D. Ollason, V. Valtchev, P. Woodland, "The HTK book - version2.2", Entropic, 1999.

17
Design of Audio-Visual Interface for Aiding Driver's Voice Commands in Automotive Environment

Kihyeon Kim, Changwon Jeon, Junho Park, Seokyeong Jeong, David K. Han, and Hanseok Ko

Abstract This chapter describes an information-modeling and integration of an embedded audio-visual speech recognition system, aimed at improving speech recognition under adverse automobile noisy environment. In particular, we employ lip-reading as an added feature for enhanced speech recognition. Lip motion feature is extracted by active shape models and the corresponding hidden Markov models are constructed for lip-reading. For realizing efficient hidden Markov models, tied-mixture technique is introduced for both visual and acoustical information. It makes the model structure simple and small while maintaining suitable recognition performance. In decoding process, the audio-visual information is integrated into the state output probabilities of hidden Markov model as multistream features. Each stream is weighted according to the signal-to-noise ratio so that the visual information becomes more dominant under adverse noisy environment of an automobile. Representative experimental results demonstrate that the audio-visual speech recognition system achieves promising performance in adverse noisy condition, making it suitable for embedded devices.

Keywords Active shape model · Audio-visual speech interface · Automatic speech recognition · Hybrid integration · Lip-reading · Mel-frequency cepstrum coefficients · Mouth model · Multistream features · SNR-dependent audio–visual information combination · Tied-mixture hidden Markov model

17.1 Introduction

Employing automatic speech recognition (ASR) system for enabling automobile information services such as voice-activated navigation system becomes a formidable challenge due to noisy automobile environment. The problem becomes particularly pronounced when the desired speech is inadvertently

H. Ko (✉)
Department of Electronics & Computer Engineering, Korea University, Seoul, Korea
e-mail: hsko@korea.ac.kr

K. Takeda et al. (eds.), *In-Vehicle Corpus and Signal Processing for Driver Behavior*, DOI 10.1007/978-0-387-79582-9_17,

contaminated by other passengers' voice or non-stationary road noises such as horns or sirens.

However, most of acoustical noise suppression approaches perform poorly under interfering noises because the noise component is estimated with the assumption that noise is stationary [1–3].

In this chapter, we present the audio–visual speech interface (AVSI) system extracting driver's audio-visual features as a solution to ensure stable recognition performance in various noisy environments. Specifically, we designed a bimodal information-modeling and integration technique which exploit lip-reading as an additional feature [4–7].

The design of the AVSI system consists of three modules. The first module handles extracting the driver's visual feature, e.g., lip shape, for visual mapping of the spoken utterances. It is obtained by employing the active shape model (ASM). The second is the acoustic module wherein features are extracted by the conventional method of Mel-frequency cepstrum coefficients (MFCCs) and SNR is estimated statistically. The third module performs information fusion by weighting the state output probabilities for each stream according to the estimated SNR. Subsequently, lower SNR assigns higher weight on the state output probabilities in visual stream.

For decoding process in the fusion module, tied-mixture hidden Markov model (TMHMM) structure is applied [8]. In this case, states use one global pool of Gaussian mixtures for each information channel. Such information modeling technique provides ability to handle the sparse data problem in training process, simply by reducing the number of Gaussians and thereby dramatically reducing the model complexity. Moreover, it can minimize the computational requirement effectively on calculation of likelihoods when it is combined with relevant ancillary refinements, such as beam control.

We demonstrate through representative experiments that the design is suitable for automobile embedded devices and produces reasonable ASR performance under severe noise conditions as well as under speech contamination by third-party utterances.

17.2 Visual Feature Extraction

In order to detect lip region, the input image is first converted from the RGB color space to the YCbCr space for its better ability of discerning features for lips. In general, Cr and Cb values of the lip region are higher than those of other regions and Cr values are relatively higher than those of Cb. For Y component, it is higher in the upper part of the lip and is lower at the corners.

In Fig. 17.1, the image A is the binary image which is extracted by applying a fixed threshold on the "MouthMap." The MouthMap, a term for face recognition, is the lip image composed of only chroma components (Cr, Cb). Image B is generated by the following equation:

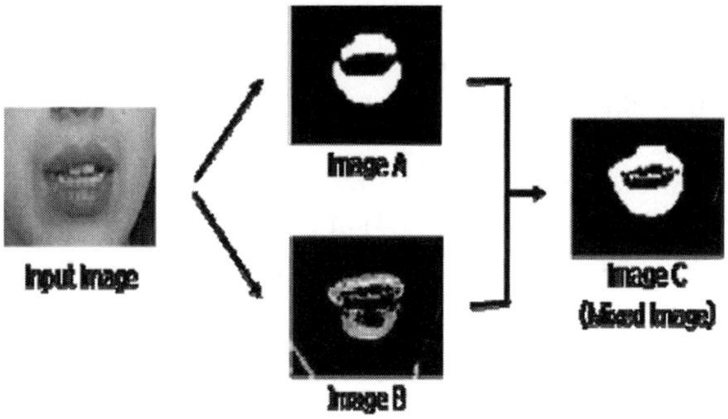

Fig. 17.1 Segmentation for lip region

$$\text{Image B} = (\max\,(\text{Cr} - \text{Y},0) - \max\,(\text{Cb} - \text{Y},0))^2. \qquad (17.1)$$

In Eq. (17.1), max(Cr−Y,0) and max(Cb−Y,0) eliminate most of the face region except the lip because chroma values are relatively higher in lip region than luminance values. The difference between the two terms emphasizes the lip part by considering the relative Cr component to Cb. As a result, the image A represents the middle region of lip well and the image B captures the edges of the lip's upper part and the corners. These two images are mixed to the binary image C for detection of the lip region.

ASM is applied to extract lip points from the detected images as shown in Fig. 17.2. Two models for the open and closed mouth states are made by mean shape models using principal component analysis (PCA). Open mouth model consists of 16 points representing the inside and the outside parts of lip, and closed mouth model consists of 10 points for only the outside part of lip as shown in Fig. 17.2 [9].

ASM forms the lip model equation as

$$X = \bar{X} + Pb, \qquad (17.2)$$

where X is a lip model, $\bar{X}$ is the mean shape model, P is a matrix of eigenvectors by PCA and b is a parameter for the variation of the mouth shape.

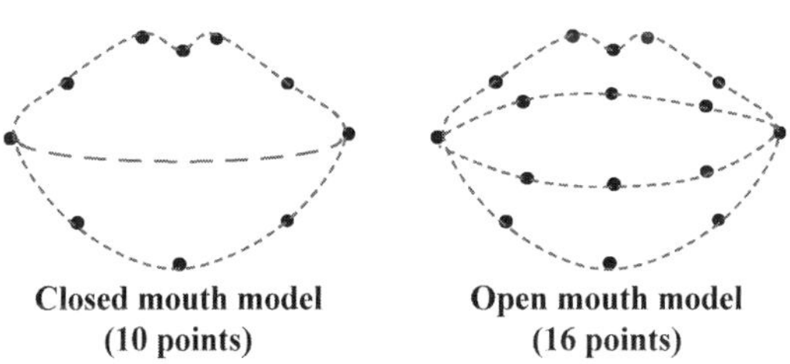

Fig. 17.2 Mean shape model (closed mouth and open mouth)

Because the lip image is binary, by counting black pixels inside the detected lip, it can be easily determined whether the mouth state is "open" or "closed." In Eq. (17.2), the appropriate mean shape model is applied according to the mouth state.

In general, ASM is an iterative method to converge to the minimum of the maximum energy defined by Eq. (17.4) at each point until the estimated position of the lip model is well aligned to the detected position of the input image. The main criterion is given by

$$\mathrm{d}\hat{b} = \arg\min_{b+\mathrm{d}b} E(\mathrm{d}x, \mathrm{d}b), \tag{17.3}$$

where E represents the energy function defined by

$$E = (F(\mathrm{d}x, \mathrm{d}b))^{\mathrm{T}} W(F(\mathrm{d}x, \mathrm{d}b)), \tag{17.4}$$

$$F(\mathrm{d}x, \mathrm{d}b) = ((X + \mathrm{d}x) - (\bar{X} + P(b + \mathrm{d}b))) \tag{17.5}$$

and W is an arbitrary cost matrix to maximize Eq. (17.4).

For the AVSI system, the visual features are constructed from the detected lip points (as illustrated in Fig. 17.3). The visual stream consists of the width and height of the mouth and the four selected points that are normalized by mean subtraction to attenuate the effect of individual lip size and to ensure the robustness as shown in Fig. 17.4.

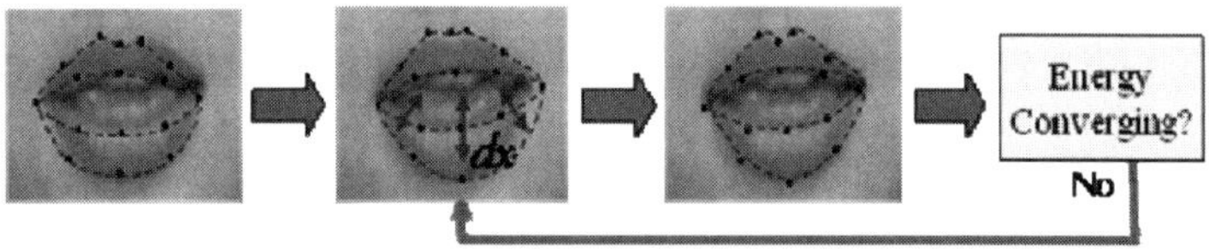

Fig. 17.3 Tracking lip shape for ASM

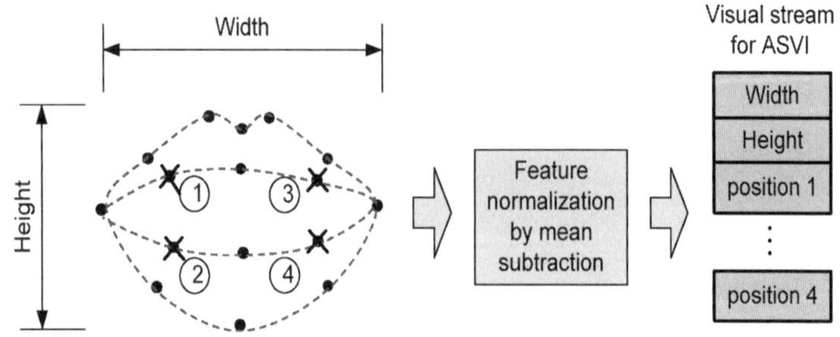

Fig. 17.4 Construction of visual feature stream for AVSI system

17.3 SNR-Dependent Audio-Visual Information Combination

After the visual feature is extracted by the ASM, linear interpolation on visual feature is employed to synchronize with the acoustic features. These features then form the multistreams for the decoding process. The acoustic and fusion modules are described in the following sections.

17.3.1 Acoustic Feature Extraction and Estimation of SNR

In the acoustic module, conventional MFCC method is applied to extract features from the audio signal as shown in Fig. 17.5.

The signal-to-noise ration (SNR) ξ is estimated by a simple spectral subtraction algorithm as

$$\xi = \alpha \cdot 10\left[\log 10\left[\frac{1}{L}\sum_{l\in L}\sum_{\text{all}k}[\Phi_Z(k,l) - \Phi_N(k)]\right] - \log 10\left[\sum_{\text{all}k}\Phi_N(k)\right]\right], \quad (17.6)$$

where k, l are the frequency and frame index; α is the compensation factor; L is the interval where the desired speech signal is activated; Φ_Z, Φ_N are power spectral densities for the input signal and the estimated noise, respectively.

The noise power spectrum is estimated by averaging audio signals in the input stream where the desired speech signal is not expected to be present. The factor α is used to compensate the estimated SNR when the noise becomes significant including cases such as non-stationary components like interfering speech or sirens. Thus, when α is set to be lower than one, the AVSI system would place more weights on the visual stream. It has been observed that the estimation procedure does not need to define exact values of α. Instead, discretely assigned values are often sufficient for the weights. Table 17.1 shows the apparent optimal audio-visual weights to ensure successful recognition performance of the AVSI system.

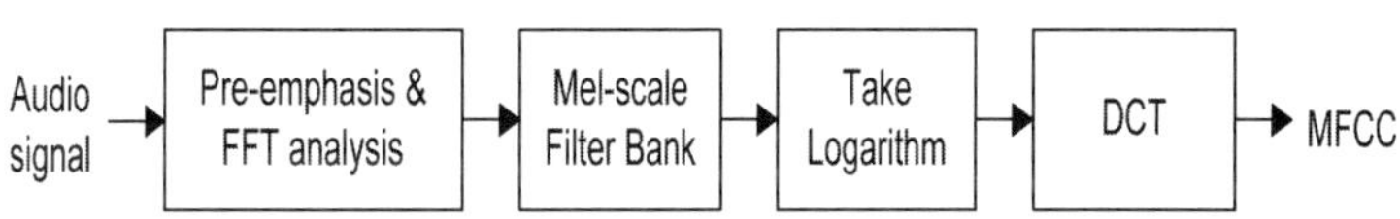

Fig. 17.5 Schematic diagram of MFCC feature extraction

Table 17.1 Optimal audio–visual weights according to various SNR levels

SNR (dB)	SNR<0	0<SNR<5	5<SNR<10	SNR>10
A-V weights	(0, 1)	(0.6, 0.4)	(0.8, 0.2)	(0.9, 0.1)

A-V audio–visual pair

Under adverse acoustic noise conditions, the visual weight becomes higher than the acoustic one since the visual information is relatively more reliable than the acoustic information.

17.3.2 Audio–Visual Model Combination

Generally, three types of structures are investigated to integrate audio-visual information: early integration, late integration, and hybrid integration [5]. However, most of the recent work has focused on the hybrid integration method due to its flexibility and robustness. This chapter develops the AVSI system based on the hybrid integration technique which is also known as multistream [4, 6]. We introduce the TMHMM as the model structure suitable for embedded applications such as telematics [8]. Figure 17.6 describes the overall process of the proposed multistream integration.

In the fusion stage, state output probability of HMM is given by

$$P(o_t|S) = P(o_{a,t}|s_a)^{\lambda_a} P(o_{v,t}|s_v)^{\lambda_v}, \tag{17.7}$$

where $O_{a,t}$, $O_{v,t}$ are the audio and visual feature streams at a frame t; S is the TMHMM state which is composed of substates S_a for the audio stream and S_v for the visual stream; λ_a, λ_v are weighting factors that are constrained by relations $\lambda_a, \lambda_v > 0$ and $\lambda_a + \lambda_v = 1$, respectively. That is, the state output probability is the linear combination of emission probabilities from corresponding substates in log-domain.

This audio-visual fusion technique has several advantages in training and decoding. First, it requires the searching process only once and thus would reduce

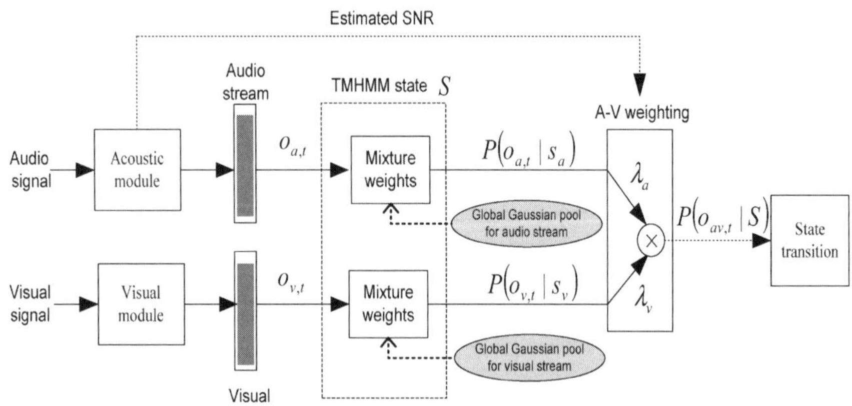

Fig. 17.6 Block diagram of TMHMM-based multistream audio-visual integration

computational requirements significantly. This is due to the fact that the emission probabilities are combined within one state before the state transition occurs, and therefore it prevents any parallel search. Although several previous works have shown that the audio-visual state propagation scheme [5] can enhance the recognition performance, improvements obtained were not significant enough for the required increase in the computational load in decoding. The second advantage is that the audio and the visual models can be trained independently if the state transition probability relies only on the acoustic model.

As mentioned earlier, the state output probability is composed of simple audio and visual emission probabilities. In other words, the substates S_a and S_v do not affect each other in decoding and can be trained independently. Moreover, the introduction of TMHMM makes the model structure simple by having to construct just one global Gaussian pool for each information channel. In this way, every state holds the pools in common and has only mixture weights. Hence, for training, the data sparse problem can be handled simply by decreasing the number of Gaussian components in the pool while the computational load for the decoding can be reduced by the beam pruning according to the mixture weights. Finally, weights for audio and visual streams are determined automatically by the estimated SNR in Section 17.3.1. Therefore, the system can adapt to the variation of the environmental noise.

These advantages ensure that the audio-visual speech recognition system has reasonable performances and can be easily implemented for automobile embedded devices.

17.4 Experiments and Results

For experiments, acoustic models were synthesized by using a decision tree over previously well-trained TMHMMs. Visual models were trained by real video data and formed a global Gaussian pool. State transition probabilities rely on the acoustic model. As test data, we used the audio-visual DB which was composed of recording by 20 Korean speakers. Each speaker pronounced 20 arbitrarily selected Korean words twice. The total number of test files was 800. Automobile noises including the third utterance were added to the clean audio data with various ranges of SNR from −10 to 20 dB.

Figure 17.7 shows word error rates (WERs) of the proposed audio-visual speech recognition system in comparison with the baseline recognizer using audio information only. The conventional ASR system naturally has higher WERs at lower SNR values. The performance falls rapidly below 10 dB of SNR. On the other hand, the proposed audio-visual speech interface system ensures the recognition performance under 43% of WER even when the input SNR is lower than 0 dB as well as gives improved results constantly in overall SNR range in comparison with the conventional system.

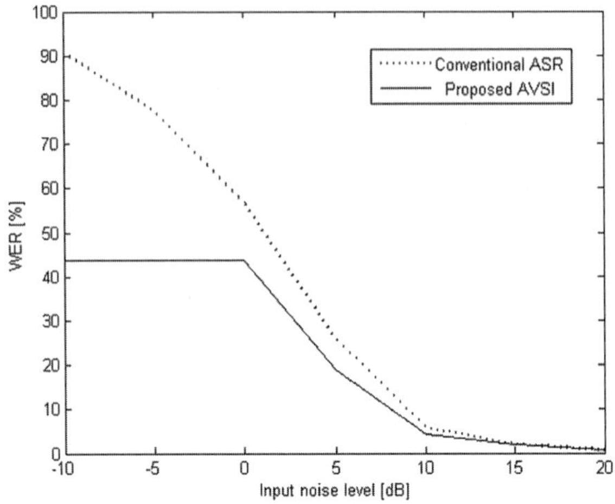

Fig. 17.7 WER of the proposed audio-visual speech recognition system in comparison with conventional ASR system

17.5 Conclusion

This chapter describes the implementation of an audio-visual speech recognition system designed to ensure the performance in adverse noise environment. The system uses multistream and TMHMM structure suitable for automobile embedded devices. The approach allows the bimodal fusion to be simple while maintaining deployable level of recognition performance. Moreover, the proposed SNR estimation and the automatic weighting technique enable the system to actively adapt to various noise conditions. Several experiments have shown that the AVSI system can be a reliable solution, compared to the conventional ASR system, even in noisy environment including interfering noise such as third-party speech.

Acknowledgments This research was supported by the MIC (Ministry of Information and Communication), Korea, under the ITFSIP (IT Foreign Specialist Inviting Program) supervised by the IITA (Institute of Information Technology Advancement).

References

1. S. F. Boll, "Suppression of acoustic noise in speech using spectral subtraction," IEEE Trans. Acoust. Speech Signal Process., vol. ASSP-27, pp. 113–120, Apr. 1979.
2. Y. Ephraim and D. Malah, "Speech enhancement using a minimum mean-square error log-spectral amplitude estimator," IEEE Trans. Acoust. Speech Signal Process., vol. ASSP-33, pp. 443–445, Apr. 1985.
3. S. Gannot et al., "Signal enhancement using beamforming and nonstationarity with applications to speech," IEEE Trans. Signal Process., vol. 49, pp. 1614–2001, Aug. 2001.

4. S. Dupont and J. Luettin, "Audio-visual speech modeling for continuous speech recognition," IEEE Trans. Multimedia, vol. 2, pp. 141–151, Sept. 2000.
5. G. Potamianos et al., "Recent advances in the automatic recognition of audiovisual speech," Proc. IEEE, vol. 91, pp. 1306–1326, Sept. 2003.
6. J. F. Perez et al., "Lip reading for robust speech recognition on embedded devices," IEEE International Conference on Acoustics, Speech and Signal Processing, 2005.
7. T. J. Hazen, "Visual model structures and synchrony constraints for audio–visual speech recognition," IEEE Trans. Audio Speech Lang. Process., vol. 14, pp. 1082–1089, May 2006.
8. J. Park and H. Ko, "Achieving a reliable compact acoustic model for embedded speech recognition system with high confusion frequency model handling," Speech Commun., vol. 48, pp. 737–745, June 2006.
9. A. Caplier, "Lip detection and tracking," 11th International Conference on Image Analysis and Processing, pp. 8–13, 2001.

18
Estimation of High-Variance Vehicular Noise

Bowon Lee and Mark Hasegawa-Johnson

Abstract In this chapter, we describe a method of minimum mean-squared error (MMSE) a posteriori estimation of high-variance vehicular noise. The proposed method considers spectral instances of noise as sampled values from a stochastic noise process and estimates them with given statistical properties of noise and current noisy observation. Accuracy of the noise estimation method is evaluated in terms of the accuracy of a spectrum-based voice activity detection (VAD), in which speech presence is determined by the a priori and a posteriori signal-to-noise ratios (SNRs) in each frequency bin. VAD experiments are performed on clean speech data by adding four different types of vehicular noise, each with the SNR varying from −10 to 20 dB. Also, isolated digit recognition experiments are performed using original noisy recordings from the AVICAR corpus. Experimental results show that the proposed noise estimation method outperforms both the MMSE a priori noise estimation and autoregressive noise adaptation methods especially for low SNR.

Keywords Minimum mean-squared error estimation · Noise spectrum estimation · Short-time Fourier transform · Periodogram · Voice activity detection · Automatic speech recognition · Speech enhancement · Autoregressive adaptation · Error propagation · Hidden Markov model · Word error rate

18.1 Introduction

Speech processing systems such as speech coding and automatic speech recognition are typically designed for clean speech as input. In many practical situations such as in vehicular environments, background noise is added to the input speech and is detrimental to the performance of speech processing

B. Lee (✉)
Hewlett-Packard Laboratories, Palo Alto, CA, 94304, USA
e-mail: bowon.lee@hp.com

K. Takeda et al. (eds.), *In-Vehicle Corpus and Signal Processing for Driver Behavior*, DOI 10.1007/978-0-387-79582-9_18,

systems. In order for those systems to operate as they are designed, we need to use only speech components from noisy speech. This is often referred to as speech enhancement and it requires estimation of noise spectrum. Given noisy speech with speech presence uncertainty, noise spectrum estimation is a very challenging problem.

Speech enhancement algorithms often depend on the existence of a robust voice activity detector (VAD). Even without speech enhancement, a good VAD can substantially improve the word error rate of automatic speech recognition in noise [1]. VAD can be modeled as a likelihood ratio test, evaluating the relative likelihoods of speech presence versus absence [2], or as an explicit computation of speech presence probability [3, 4]. Statistical VAD algorithms are mostly based on the signal-to-noise ratio (SNR) [5–7]. All of these methods require noise spectrum estimation, thus its accuracy is critical to their performance.

Most systems depend on the minimum mean-squared error (MMSE) estimate of the noise power spectrum, i.e., an estimate of the expected value of the noise power in each frequency bin of the short-time Fourier transform (STFT). The expected noise power may be estimated from several frames which contain only noise. Alternatively, noise power may be recursively updated with a fixed adaptation coefficient by assuming that noise is non-stationary, but slowly varying. Sohn and Sung [6] proposed an autoregressive noise adaptation method with a variable adaptation coefficient that depends on the speech presence probability, with which accurate estimation of speech presence is crucial.

With higher noise power, noise spectrum has higher variance, and therefore, even though the noise is stationary, the MMSE estimate of the noise spectrum may not be close to the noise spectrum of the current observation. High noise power also disrupts autoregressive noise adaptation methods because these methods depend on an estimate of speech presence probability: with low SNR, estimated speech presence probabilities are less accurate. For these reasons, speech enhancement and VAD algorithms that perform well with high SNR may nevertheless fail with low SNR such as in vehicular environments.

This chapter describes an MMSE *a posteriori* estimation of noise based on the MMSE *a priori* estimation of noise, combined with a current noisy observation, employing speech presence uncertainty. This method treats spectral instances of noise as sampled values of an independent and identically distributed (iid) random variable; thus, unlike noise adaptation methods, the proposed method does not assume that noise spectral amplitude is predictable from its own recent history.

Experimental results show that the proposed noise estimation achieves higher VAD accuracy as well as isolated word recognition accuracy in an automotive environment compared to MMSE *a priori* noise estimation and autoregressive noise adaptation methods.

18.2 Background

18.2.1 Statistical Noise Model

Consider a sequence of stationary noise n as a random process with an unknown probability density function (pdf) with zero mean. Let a length L short-time Fourier transform (STFT) of n of the mth frame be given by

$$N_k^m = \sum_{l=0}^{L-1} n[l + mL] \exp\left\{ -\mathrm{j}\frac{2\pi kl}{L} \right\}. \tag{18.1}$$

If we consider that each STFT coefficient of n is a weighted sum of samples of the corresponding random process, then according to the central limit theorem, as $L \to \infty$, the STFT coefficients N_k^m asymptotically have a Gaussian pdf with zero mean [3]. Thus, the pdf of the kth frequency bin, N_k^m, can be expressed as

$$p(N_k^m) = \frac{1}{\pi\lambda_N(k)} \exp\left\{ -\frac{\left|N_k^m\right|^2}{\lambda_N(k)} \right\}, \tag{18.2}$$

where $\lambda_N(k) = E\left[\left|N_k^m\right|^2\right]$ denotes the variance of a discrete Fourier transform (DFT) coefficient of noise in the kth frequency bin. Since it has a Gaussian distribution, we can show that the power at the kth spectral component $\left|N_k^m\right|^2$ has an exponential pdf with mean $\lambda_N(k)$. Thus, the variance of the DFT of noise $\lambda_N(k)$ is equivalent to the MMSE estimation of noise power.

Figure 18.1 depicts a histogram of noise power versus an exponential pdf: the solid line is a normalized histogram of noise power in the *100*th frequency bin of a 400 point DFT of a white Gaussian noise and the dashed line is an exponential pdf.

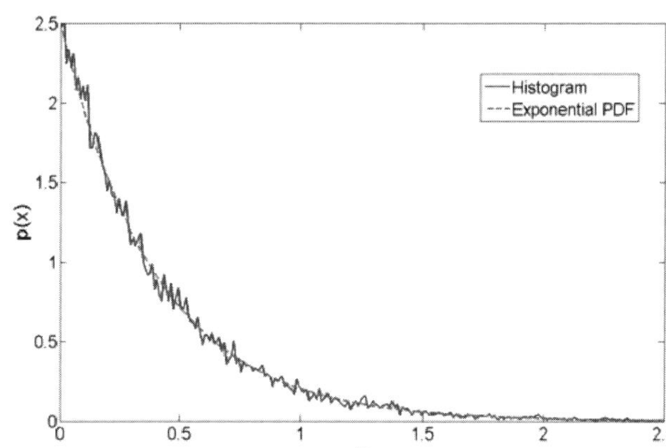

Fig. 18.1 Normalized histogram of noise power compared to an exponential pdf

18.2.2 MMSE A Priori Noise Estimation

Since infinite length noise sequence is not available for calculating the true mean $\lambda_N(k)$ of noise variance, we need to calculate its estimate $\hat{\lambda}_N(k)$. The most commonly used noise estimation method given a finite noise sequence is the periodogram estimation:

$$\hat{\lambda}_N^m(k) = \left|N_k^m\right|^2, \tag{18.3}$$

where $\left|N_k^m\right|^2$ is the STFT of the noise of the mth frame as defined in Eq. (18.1) above.

Since $\hat{\lambda}_N^m(k)$ is an exponentially distributed random variable with its standard deviation equivalent to its mean, we use the Bartlett's procedure to reduce its variance by averaging M frames:

$$\bar{\lambda}_N(k) = \frac{1}{M}\sum_{m=0}^{M-1}\hat{\lambda}_N^m(k). \tag{18.4}$$

This method requires a length LM sequence of noise-only observations. Here $\bar{\lambda}_N(k)$ is an unbiased and consistent estimator of $\lambda_N(k)$:

$$E\left[\bar{\lambda}_N(k)\right] = \lambda_N(k), \tag{18.5}$$

$$E\left[\left(\bar{\lambda}_N(k) - \lambda_N(k)\right)^2\right] = \frac{1}{M}\lambda_N(k)^2. \tag{18.6}$$

However, Eqs. (18.5) and (18.6) do not imply that $\bar{\lambda}_N(k)$ predicts any particular instance of $\left|N_k^m\right|^2$ with high accuracy: $\left|N_k^m\right|^2$ is exponentially distributed, so its standard deviation still equals its mean. With high noise power, $\bar{\lambda}_N(k)$ may not be as close to the true noise power compared to that with low noise power.

18.2.3 Noise Estimation with Speech Presence Uncertainty

So far, we have discussed noise estimation problem with noise-only observations. In practice, we have noisy speech observation x which may contain noise only, i.e., $x = n$ or speech plus noise, i.e., $x = s + n$, with S representing speech signal. The most challenging aspect of this problem is that we do not know speech presence a priori, given noisy speech observations. Considering this, we formulate two hypotheses:

$$\begin{cases} H_0 : X_k^m = N_k^m, & \text{speech absent} \\ H_1 : X_k^m = S_k^m + N_k^m, & \text{speech present} \end{cases}, \tag{18.7}$$

where S_k^m, N_k^m, and X_k^m are STFT coefficients of speech, noise, and noisy speech, respectively. The pdf of X_k^m given H_0 is equivalent to Eq. (18.2) and the pdf given H_1 is [3]

$$p(X_k|H_1) = \frac{1}{\pi(\lambda_N(k) + \lambda_S(k))} \exp\left\{ -\frac{|X_k|^2}{\lambda_N(k) + \lambda_S(k)} \right\},$$

where $\lambda_S(k) = E\left[\left|S_k^m\right|^2\right]$ and $\lambda_N(k) = E\left[\left|N_k^m\right|^2\right]$ denote the speech and noise variance, respectively, and it is assumed that speech and noise are uncorrelated. Then, the likelihood ratio at the kth frequency bin is

$$\Lambda_k = \frac{p(X_k|H_1)}{p(X_k|H_0)} = \frac{1}{1+\xi_k} \exp\left\{ \frac{\gamma_k \xi_k}{1+\xi_k} \right\}, \tag{18.8}$$

where $\xi_k = \lambda_S(k)/\lambda_S(k)\lambda_N(k)\lambda_N(k)$ and $\gamma_k = \left|X_k^m\right|^2 \Big/ \left|X_k^m\right|^2 \lambda_N(k)\lambda_N(k)$ are defined as *a priori* and *a posteriori* SNR, respectively [3]. Speech presence probability is therefore dependent upon speech and noise variance and their accurate estimation is crucial for *a posteriori* noise estimation methods, which will be discussed in the following section.

18.2.4 MMSE A Posteriori Noise Estimation

Considering the speech presence uncertainty, the MMSE estimate of the noise at the kth frequency bin in the mth frame given current noisy observation is [2]

$$\begin{aligned} \hat{\lambda}_N^m(k) =& E\left[\left|N_k^m\right|^2 \middle| X_k^m\right] \\ =& E\left[\left|N_k^m\right|^2 \middle| H_0\right] p(H_0|X_k^m) + E\left[\left|N_k^m\right|^2 \middle| H_1\right] p(H_1|X_k^m). \end{aligned} \tag{18.9}$$

Using Bayes' rule,

$$\begin{aligned} p(H_0|X_k^m) =& \frac{p(X_k^m|H_0)p(H_0)}{p(X_k^m|H_0)p(H_0) + p(X_k^m|H_1)p(H_1)} \\ =& \frac{1}{1+\varepsilon\Lambda_k^m}, \end{aligned} \tag{18.10}$$

where $\varepsilon = p(H_1)/p(H_1)p(H_0)p(H_0)$ and $\Lambda_k^m = p(X_k^m|H_1)/p(X_k^m|H_1)p(X_k^m|H_0)$ $p(X_k^m|H_0)$ is the likelihood ratio of speech presence of the mth frame. We can derive $p(H_1|X_k^m)$ similarly,

$$p(H_1|X_k^m) = \frac{\varepsilon\Lambda_k^m}{1+\varepsilon\Lambda_k^m}. \tag{18.11}$$

If we let Eq. $\beta_k^m = p(H_1|X_k^m) = \varepsilon\Lambda_k^m/\varepsilon\Lambda_k^m(1+\varepsilon\Lambda_k^m)(1+\varepsilon\Lambda_k^m)$ and substitute Eqs. (18.10) and (18.11) into (18.9), then

$$\hat{\lambda}_N^m(k) = \beta_k^m E\Big[|N_k^m|^2|H_1\Big] + (1-\beta_k^m)E\Big[|N_k^m|^2|H_0\Big]. \tag{18.12}$$

Therefore, proper estimation of β_k^m representing speech presence probability as well as expectation of noise power in each hypothesis provides the MMSE a posteriori estimation of noise spectrum.

18.2.5 Autoregressive Noise Adaptation

In Eq. (18.12), we need the estimates of noise spectrum of each hypothesis, $E\Big[|N_k^m|^2|H_0\Big]$ and $E\Big[|N_k^m|^2|H_1\Big]$. Under the hypothesis H_0, we can use the current noisy observation [2, 6]:

$$E\Big[|N_k^m|^2|H_0\Big] = |X_k^m|^2. \tag{18.13}$$

Under the hypothesis H_1, $|X_k^m|^2$ contains speech as well as noise and is therefore not an accurate estimate of the noise power. Sohn and Sung [6] proposed that assuming that the likelihood ratio Λ_k^m has been correctly estimated in all previous frames, the best available estimate of the noise power is

$$E\Big[|N_k^m|^2|H_1\Big] = \hat{\lambda}_N^{m-1}(k). \tag{18.14}$$

Combining Eqs. (18.12) through (18.14) yields

$$\hat{\lambda}_N^m(k) = \beta_k^m\hat{\lambda}_N^{m-1}(k) + (1-\beta_k^m)|X_k^m|^2. \tag{18.15}$$

It is also proposed that, if Λ_k^m is an accurate estimate of the speech presence probability in each frame, then Eq. (18.15) is an equally accurate estimate of the noise power in the mth frame [6]. Under these circumstances, $\hat{\lambda}_N^m(k)$ takes into account all information about the underlying noise process that can be extracted from frames up to and including the current frame.

18.3 Estimation of High-Variance Noise

18.3.1 Speech Presence Probability with Low SNR

The autoregressive noise estimator $\hat{\lambda}_N^m(k)$ proposed in Eq. (18.15) is optimal, if and only if the speech presence probability estimate β_k^m is accurate. Unfortunately, with low SNR, β_k^m is itself a random variable with high variance. β_k^m is a sigmoid transformation of the random variable $|X_k^m|^2$:

$$\beta_k^m = \frac{e^{|X_k^m|^2/(a_k\lambda_N(k))}}{(a_k/a_k\varepsilon\varepsilon) + e^{|X_k^m|^2/(a_k\lambda_N(k))}}$$

where $a_k = (1+\xi_k)/(1+\xi_k)\xi_k\xi_k$. The input threshold of the sigmoid—the value of $|X_k^m|^2$ at which $\beta_k^m = 0.5$—is given by $\theta_k = a_k\lambda_N(k)\log(a_k/a_k\varepsilon\varepsilon)$. Any noise-only frame in which $|N_k^m|^2 > \theta_k$ will cause a "false positive": $\beta_k^m \approx 1$ despite the absence of speech. Equation (18.15) prohibits these false positives from contributing to the autoregressive estimate$\hat{\lambda}_N^m(k)$; therefore, over time, the estimate $\hat{\lambda}_N^m(k)$ tends to underestimate the true expected value $E\left[|N_k^m|^2\right]$ and to overestimate the probability of speech presence in any given frame.

In order to more precisely estimate the amount by which autoregressive noise adaptation underestimates $\lambda_N(k)$ in low-SNR environments, let us treat β_k^m as a binary random variable— a unit step function of $|N_k^m|^2$, rather than a sigmoid function. Define $\rho = P(\beta_k^m \geq 0.5)$; under the assumption, $E[\beta_k^m] = \rho$. By integrating the pdf of $|N_k^m|^2$, we find that

$$\rho = \int_{a_k\log(a_k/a_k\varepsilon\varepsilon)}^{\infty} e^{-t}dt = \left(\frac{a_k}{\varepsilon}\right)^{-a_k}. \tag{18.17}$$

In terms of ρ, the expected value of $\hat{\lambda}_N^m(k)$ is approximately

$$E\left[\hat{\lambda}_N^m(k)\right] \approx \rho E\left[\hat{\lambda}_N^{m-1}(k)\right] + (1-\rho)E\left[|X_k^m|^2 | \beta_k^m < 0.5\right].$$

However,

$$\begin{aligned}(1-\rho)E\left[|X_k^m|^2|\beta_k^m<0.5\right] \\ = \lambda_N(k)\int_0^{a_k\log(a_k/a_k\varepsilon\varepsilon)} te^{-t}dt \\ = \lambda_N(k)[1-\rho-\log\rho].\end{aligned} \tag{18.18}$$

Combining equations, we find that

$$E\left[\hat{\lambda}_N^m(k)\right] \approx \rho\hat{\lambda}_N^{m-1}(k) + \lambda_N(k)[1-\rho-\log\rho]. \tag{18.19}$$

If we begin with a perfect estimate $\hat{\lambda}_N^1(k) = \lambda_N(k)$, then Eq. (18.19) demonstrates that $\hat{\lambda}_N^m(k)$ will tend to decay over time, with an initial slope of $\rho\log\rho\lambda_N(k) < 0$.

The scaling factor $\rho\log\rho$ is most negative at values of $\rho = e^{-1}$; for example, if $\varepsilon = 1$, then the smallest value of $\rho\log\rho$ (and therefore, according to the estimate in Eq. (18.19), the worst underestimation of $\lambda_N(k)$ by the autoregressive estimator) occurs at an SNR of $\xi_k = 1.3$, quite close to 0 dB SNR.

18.3.2 Proposed Noise Estimation Method

In high-noise environments with low SNR, error propagation in Eq. (18.15) is an important problem. Error propagation can be avoided by applying a certain amount of prior knowledge to the problem. For example, if the noise process is known to be stationary and if the first M frames of the signal are known to contain no speech, then an *a priori* periodogram estimate $\bar{\lambda}_N(k)$ of $E\left[|N_k^m|^2\right]$ with known standard error may be computed using Eq. (18.4). If we assume that intervening frames provide no further information about $E\left[|N_k^m|^2\right]$, then

$$E\left[|N_k^m|^2|H_1\right] = \bar{\lambda}_N(k) \tag{18.20}$$

and Eq. (18.12) becomes

$$\hat{\lambda}_N^m(k) = \beta_k^m \bar{\lambda}_N(k) + \left(1 - \beta_k^m\right)\left|X_k^m\right|^2. \tag{18.21}$$

If it is highly likely that the speech is present, i.e., $\Lambda_k^m >> 1$, thus $\beta_k^m \approx 1$, then Eq. (18.21) sets $\hat{\lambda}_N^m(k)$ equal to the mean estimate of the noise spectrum. Thus the method proposed in Eq. (18.21) is subject to false-positive errors, just like the autoregressive estimator, but Eq. (18.21) does not propagate error. Instead, a false-positive frame is treated just like any other frame about which we have no certain knowledge of the noise spectrum: the noise estimate is backed off to the a priori noise estimator $\bar{\lambda}_N(k)$.

The proposed noise spectrum estimation method can be interpreted as an a posteriori MMSE estimate of the noise power in the current frame, when the noise process is stationary but with high variance. Experimental results show that the proposed noise estimation method provides higher accuracy especially for low-SNR cases.

18.4 Experiments

Noise estimation methods were evaluated using two experimental tests. First, noise and speech were electronically mixed, and the noise estimation methods in Eqs. (18.4), (18.15), and (18.21) were tested in the task of voice activity detection (VAD). Second, original noisy speech data were end-pointed using VAD with

Table 18.1 Four noise conditions from the AVICAR database

Noise condition	Description
35U	Car running at 35 mph with windows closed
35D	Car running at 35 mph with windows open
55U	Car running at 55 mph with windows closed
55D	Car running at 55 mph with windows open

each of the three noise estimation methods, and word error rates (WER) were computed using mixture Gaussian HMMs.

VAD tests used 62 sentences from the TIMIT database [8]. One second of silence was inserted between adjacent sentences, making a total duration of 212s, of which 54% seconds contain speech. Frame duration chosen for experiments is 10 ms and each frame is marked as either speech or silence according to the transcription by marking frames with speech in more than 50% of their duration to be speech, and silence otherwise.

We have included four different car noises out of five from the AVICAR database [9] by increasing the SNRs from −10 to 20 dB with 5 dB increments for each noise condition. (See Table 18.1.) AVICAR is a database of multi-camera, multi-microphone audiovisual speech acquired from 100 talkers in moving cars (www.ifp.uiuc.edu/speech/AVICAR/).

The best audio-only isolated digit WER achieved on this corpus, averaged across all noise conditions, is 3.95%, using a missing-features approach called the phoneme restoration HMM [1]. The best video-only isolated digit WER on this corpus is 62.5% [10]. Audiovisual error rates have not improved upon audio-only rates for this corpus.

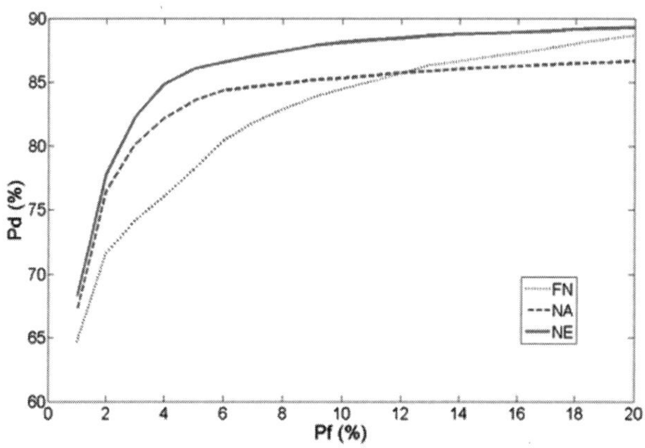

Fig. 18.2 ROC curves for 5 dB SNR with 55U noise condition, *FN* fixed noise, *NA* autoregressive noise adaptation, *NE* proposed noise estimation

Table 18.2 Summary of P_ds of the VAD experiments ($P_f = 5\%$)

	NE	NA	FN
SNR	P_d (%)	P_d (%)	P_d (%)
−10 dB	35.89	25.76	28.64
−5 dB	55.41	45.65	47.31
0 dB	72.73	65.69	64.76
5 dB	84.11	81.12	77.88
10 dB	90.74	89.83	86.99
15 dB	93.47	92.90	91.91
20 dB	94.28	94.14	94.09
Overall	75.23	70.73	70.22

FN fixed noise, *NA* autoregressive noise adaptation, *NE* proposed noise estimation

Descriptions of the noise conditions extracted from the database are listed in Table 18.1. We used the initial 20 frames (200 ms) for the mean estimate of noise assuming that they contain only noise.

Performance of VAD with different noise estimation methods is compared by correct speech detection and false alarm probabilities (P_d and P_f). Figure 18.2 depicts the receiver operation characteristics (ROC) for 5 dB SNR with noise condition 55U for three different noise estimation methods. "Fixed noise" estimation (FN) refers to Eq. (18.4); "noise adaptation" (NA) refers to Eq. (18.15); "noise estimation" (NE) refers to the proposed method, Eq. (18.21). In order to quantitatively present the accuracies of different noise estimation methods, we set the VAD threshold such that $P_f = 5\%$ and compared P_d for each noise estimation method. A summary of results across all noise conditions according to the SNRs is in Table 18.2. We can see from Table 18.2 that the autoregressive noise adaptation method performs worse than the fixed noise spectrum when the SNR is lower than 0 dB, which illustrates that with low SNR, the noise estimate in the previous frames are significantly different from the current noise spectrum, thus it makes VAD perform worse. The VAD with the proposed noise estimation method has higher accuracy compared to the other two noise estimation methods especially for low SNR.

The three VADs described in Fig. 18.2 and Table 18.2 were used to endpoint isolated digit utterances prior to automatic speech recognition (ASR). In this experiment, all audio signals are original recordings from the AVICAR corpus, therefore "ground truth" for the VAD is unknown (there is no ground truth specification of the beginning or ending of speech in each file). ASR was conducted using mixture Gaussian hidden Markov models (HMMs). HMMs were trained on data from 60 talkers (all noise conditions), the number of Gaussians was increased until error rate reached a minimum on data from 20 talkers, and then error rate was computed for the remaining 20 talkers. Fivefold cross-validation was used to compute the WER in Table 18.3. WERs reported

Table 18.3 WER in percent, HMM isolated digit recognizers, noisy speech AVICAR recordings, fivefold cross-validation

Noise condition	Baseline	NE	NA	FN
35U	2.30	2.80	2.86	3.47
35D	2.40	3.64	3.75	4.34
55U	3.51	4.32	4.9	5.38
55D	6.02	7.00	8.41	9.21
Overall	3.49	4.07	4.47	5.04

for each noise condition summarize 2000 tokens each, with a maximum WER of about 5%, therefore WER differences of 0.8% are significant at the $p = 0.05$ level. WERs reported in the final row of each table summarize 10,000 tokens, therefore WER differences of 0.35% are significant. All audio signals are the result of delay-and-sum beamforming using a seven-microphone horizontal array.

18.5 Conclusion

In this chapter, we described a MMSE a posteriori noise estimation method by considering the instantaneous noise spectrum as sampled values from an underlying noise process. Experimental results show that the proposed noise estimation method provides higher accuracy for VAD and for isolated digit recognition than the MMSE a priori estimate or autoregressive noise adaptation methods with high-variance vehicular noise.

Acknowledgments Experiments described in this chapter were supported by the Motorola Corporation and by NSF grant number IIS 05-34106. Manuscript preparation funds were provided by the Hewlett-Packard Company. Opinions, results, and conclusions are those of the authors and do not necessarily represent the opinions and conclusions of Hewlett-Packard, Motorola, or the National Science Foundation.

References

1. B. Lee, Robust speech recognition in a car using a microphone array, Ph.D. thesis, University of Illinois at Urbana-Champaign, 2006.
2. R. J. McAulay and M. L. Malpass, “Speech enhancement using a soft-decision noise suppression filter,” IEEE Trans. Acoust., Speech Signal Process., vol. 28, pp. 137–145, 1980.
3. Y. Ephraim and D. Malah, “Speech enhancement using a minimum mean-square error short-time spectral amplitude estimator,” IEEE Trans. Acoust., Speech Signal Process., vol. 32, pp. 1109–1121, 1984.
4. I.-Y. Soon, S.-N. Koh, and C.-K. Yeo, “Improved noise suppression filter using self-adaptive estimator of probability of speech absence,” Signal Process., vol. 75, no. 2, pp. 151–159, 1999.

5. J. Sohn, N.-S. Kim, and W. Sung, "A statistical model-based voice activity detection," IEEE Signal Process. Lett., vol. 6, no. 1, pp. 1–3, 1999.
6. J. Sohn and W. Sung, "A voice activity detector employing soft decision based noise spectrum adaptation," Proc. Int. Conf. Acoust. Speech and Signal Process., pp. 365–368, 1998.
7. Y. D. Cho and A. Kondoz, "Analysis and improvement of a statistical model-based voice activity detector," IEEE Signal Process. Lett., vol. 8, no. 10, pp. 276–278, 2001.
8. V. Zue, S. Seneff, and J. Glass, "Speech database development at MIT: TIMIT and beyond," Speech Commun, vol. 9, no. 4, pp. 351–356, 1990.
9. B. Lee, M. Hasegawa-Johnson, C. Goudeseune, S. Kamdar, S. Borys, M. Liu, and T. Huang, "AVICAR: audiovisual speech corpus in a car environment," Proc. Int. Conf. Spoken Lang. Process., 2004.
10. Y. Fu, X. Zhou, M. Liu, M. Hasegawa-Johnson, and T. Huang, "Lipreading by locality discriminant graph," Proc. ICIP, 2007.

19 Feature Compensation Employing Model Combination for Robust In-Vehicle Speech Recognition

Wooil Kim and John H. L. Hansen

Abstract An effective feature compensation method is evaluated for reliable speech recognition in real-life in-vehicle environments. The CU-Move corpus, previously collected by RSPG (currently, CRSS) (http://www.utdallas.edu/research/utdrive; Hansen et al., *DSP for In-Vehicle and Mobile Systems*, 2004), contains a range of speech and noise signals collected for a number of speakers from across the United States under actual driving conditions. PCGMM (parallel combined Gaussian mixture model)-based feature compensation (Kim et al., *Eurospeech 2003*, 2003; Kim et al., *ICASSP 2004*, 2004), considered in this chapter, utilizes parallel model combination to generate noise-corrupted speech models by combining clean speech and noise models. In order to address unknown time-varying background noise, an interpolation method of multiple environmental models is employed. To alleviate computational expenses due to multiple models, a noise transition model is proposed, which is motivated from the noise language modeling concept developed in Environmental Sniffing (Akbacak and Hansen, *IEEE Trans Audio Speech Lang Process*, 15(2): 465–477, 2007). The PCGMM method and the proposed scheme are evaluated on the connected single digits portion of the CU-Move database using the Aurora2 evaluation toolkit. Experimental results indicate that our feature compensation method is effective for improving speech recognition in real-life in-vehicle conditions. Here, a 26.78% computational reduction was obtained by employing the noise transition model with only a slight change in overall recognition performance. The resulting system therefore demonstrates an effective speech recognition strategy for robust speech recognition for noisy in-vehicle environments.

Keywords Speech recognition · In-vehicle Condition · Feature compensation · Model combination · Noise transition model · Route Navigation Dialog · CU-Move · UTDrive

J.H.L. Hansen (✉)
Center for Robust Speech Systems (CRSS), Erik Jonsson School of Engineering and Computer Science, University of Texas at Dallas, Richardson, Texas, USA
e-mail: John.Hansen@utdallas.edu

K. Takeda et al. (eds.), *In-Vehicle Corpus and Signal Processing for Driver Behavior*, DOI 10.1007/978-0-387-79582-9_19,

19.1 Introduction

Acoustic difference between training environments and conditions where actual speech recognition systems operate is one of the primary factors that degrade speech recognition accuracy, and the presence of time-varying background noise is generally the primary factor. This is especially true for in-vehicle speech systems which face the problem of robust speech recognition in order to address a range of severe changing background noise conditions. The problem is compounded since speech recognition in the car is typically needed for hands-free command and control or for interactive dialog systems for information across such as route navigation. Both require near real-time speech recognition performance to be effective for automotive applications.

This chapter investigates the performance of feature compensation schemes in a real-life in-vehicle environment, with the goal of achieving low complexity in computation. The CU-Move corpus has been built to develop reliable speech systems for in-vehicle domains and contains a range of acoustic scenarios expected to be observed during real-life car driving [1,2]. The corpus was used for research in multi-sensor array processing for noise suppression and speech recognition in cars [3]. Therefore, performance evaluation on CU-Move database can indicate the reliability and effectiveness of the targeted algorithm in actual in-vehicle conditions.

In this study, our previously proposed PCGMM (parallel combined Gaussian mixture model)-based feature compensation method [4] is considered as a solution to address the background noise of in-vehicle conditions. PCGMM-based method employs model combination to obtain a noise-corrupted speech model and operates in the cepstral domain. By using model combination, the PCGMM scheme eliminates the prior training which requires a noise-corrupted speech database, which is an absolute requirement in conventional data-driven methods. Independent access to the noise model makes its adaptation in the non-speech interval possible. An interpolation method employing multiple environmental noise models was also developed to address unknown or time-varying noise conditions [5]. In order to reduce the computational expense due to the use of multiple models, we propose to employ a noise transition model for the multi-model approach, which is motivated from noise language modeling in our previous work [6].

This chapter is organized as follows. We first review the CU-Move corpus used for this study in Section 19.2. In Section 19.3, the PCGMM-based feature compensation method employed in our work will be discussed followed by the multi-model approach for PCGMM method in Section 19.4. We also discuss the noise transition model strategy in Section 19.5. Representative experimental procedures and their results are presented and discussed in Section 19.6. Finally, in Section 19.7, we have conclusions.

19.2 CU-Move Corpus

The focus of the CU-Move project [1][2] was to develop robust speech recognition algorithms for reliable car navigation employing a mixed-initiative dialog. This requires robust speech recognition across changing acoustic conditions. The CU-Move database was collected by RSPG (now CRSS at UTD), and consists of five parts: (i) command and control words; (ii) digit strings of telephone and credit numbers; (iii) street names and addresses; (iv) phonetically balanced sentences; and (v) Wizard of Oz interactive navigation conversations. A total of 500 speakers, balanced across gender and age, produced over 600 GB of data during a 6-month collection effort across the United States. The database and noise conditions are discussed in detail in [2]. We point out that the noise conditions are changing with time and reflect a range of SNR, stationarity, and spectral structure. The challenge in addressing these noise conditions is that they might be changing depending on the car being used and traffic and road conditions. In this study, we selected 10 speakers from approximately 100 speakers in Minneapolis, MN (i.e., Release 1.1A), and employed the connected single digits portion that contains speech under a range of varying complex in-vehicle noise events/conditions.

19.3 PCGMM-Based Feature Compensation

The PCGMM-based feature compensation method is based on the speech model. The distribution of the clean speech feature x in the cepstral domain is represented with a Gaussian mixture model (GMM) consisting of K components as follows:

$$p(x) = \sum_{k=1}^{K} \omega_k N(x; \mu_{x,k}, \Sigma_{x,k}). \quad (19.1)$$

It is assumed that the noisy environment degrades the GMM by moving the means and covariance matrices of the clean speech model, and the distribution of the noisy speech y can then be expressed as

$$p(y) = \sum_{k=1}^{K} \omega_k N(y; \mu_{y,k}, \Sigma_{y,k}). \quad (19.2)$$

In the PCGMM-based method, the parameters of the noise-corrupted speech model $\mu_{y,k}$ and $\Sigma_{y,k}$ are obtained through parallel model combination (PMC) [7, 8] procedure using clean speech and noise models independently [4]. It is also assumed that there is a constant bias transformation of the mean parameters of the clean speech model in the cepstral domain under the additive noisy

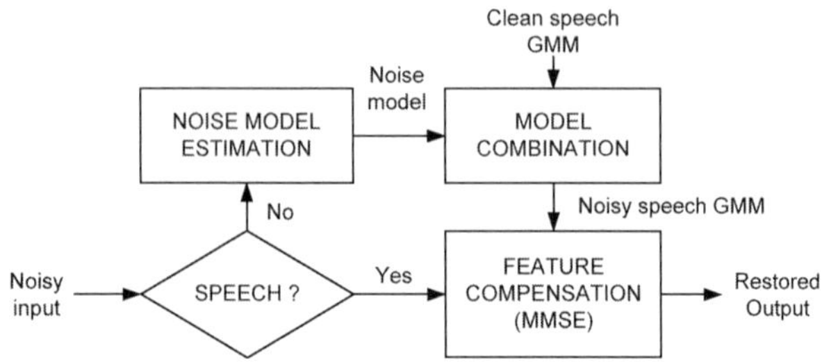

Fig. 19.1 Block diagram of PCGMM-based feature compensation method

environment, which is the assumption generally taken by other data-driven methods as follows:

$$\mu_{y,k} = \mu_{x,k} + r_k, \tag{19.3}$$

where the bias term r_k is used for reconstruction of the speech features. The MMSE equation for reconstruction of the clean speech is approximated with Eq. (4) in a manner similar to [9]:

$$\hat{x}_{\mathrm{MMSE}} = \int_X x p(x|y) \mathrm{d}x \cong y - \sum_{k=1}^{K} r_k p(k|y). \tag{19.4}$$

The posterior probability $p(k \mid y)$ can be calculated using the parameters of the noisy speech GMM $\{\omega_k, \mu_{y,k}, \sum_{x,k}\}$. Figure 19.1 presents the resulting block diagram of the PCGMM-based approach as described here.

At this point, the distinguishing properties of the PCGMM-based method are considered and compared with prior techniques. First, our method does not require an additional training procedure using a noise-corrupted speech database. After obtaining the estimated noise model from the available noise samples, the distribution model of the noise-corrupted speech can be generated via the model combination procedure. This results in a compensation method without the need of prior training data as seen in existing data-driven methods.

In the PCGMM method, estimation of the GMMs for clean speech, noise, and noisy speech as well as the reconstruction procedure is accomplished in the cepstral domain. The number of cepstral coefficients is generally smaller than for log-spectral coefficients, therefore, our method has the explicit advantage of a lower dimensional space (e.g., reduced computation). In particular, the cepstral coefficients have reduced correlation with each other as compared to the same coefficients in the log-spectral domain; therefore it is reasonable to employ diagonal covariance matrices for the GMMs in representing the models. The movement from a full covariance matrix needed for the log-spectral domain to a diagonal covariance matrix in the cepstral domain has a major reduction in both computational costs and input training data requirements for more accurate model estimation.

19.4 PCGMM-Based Method Employing Multiple Environmental Models

In the PCGMM-based method, model adaptation can be applied in order to address the time-varying background noise. In such a framework, the noise model is updated during silence periods via adaptation followed by combination of models, which again more accurately reflects the true noise for the GMM of the noisy speech. Such a framework, however, requires an accurate algorithm for silence detection and also needs considerable computational resources due to the conversion between the linear spectrum, log spectrum and cepstral domains. Therefore, applying a model adaptation technique for the noise model may not be appropriate for small resource systems such as PDAs, navigation devices and other mobile systems. In this section, we consider the PCGMM-based method which employs a combination of multiple environmental models for low resource-based ASR applications.

Utilizing multiple models estimated off-line can be effective for compensating input features adaptively under time-varying noisy conditions, thus eliminating the need for additional silence detection and online model combination. In a multiple-model method, the posterior probability of each possible environment is estimated over the incoming noisy speech. In our work, the feature reconstruction procedure is modified using a frame-by-frame formulation for real-time processing by defining the sequential posterior probability of the environment [5]. Given the incoming noisy speech feature vectors $Y_t = [y_1, y_2, \ldots, y_t]^T$, the sequential posterior probability of a specific environment GMM G_i among E models over the input speech feature Y_t can be re-written as

$$p(G_i|Y_t) = \frac{P(G_i)p(Y_{t-1}|G_i)p(y_t|G_i)}{\sum_{e=1}^{E} P(G_e)p(Y_{t-1}|G_e)p(y_t|G_e)}, \tag{19.5}$$

where $p(Y_{t-1}|G_i) = \prod_{\tau=1}^{t-1} p(y_\tau|G_i)$ and $P(G_i$ is the a priori probability of each environment i, represented as a GMM. Based on Eq. (19.4), the clean feature at frame t is reconstructed using the interpolated compensation terms from Eq. (19.5) as follows:

$$\hat{x}_{t,\text{MMSE}} \cong y_t - \sum_{e=1}^{E} p(G_e|Y_t) \sum_{k=1}^{K} r_{e,k} p(k|G_e, y_t), \tag{19.6}$$

where $r_{e,\,k}$ is a constant bias term from the kth Gaussian component of the eth environment model and $p(k \mid G_e, y_t)$ is the posterior probability for environment G_e.

When the background noise is from an environment where the number of unique types is finite, such as for in-vehicle conditions (e.g., engine noise, wind noise, turn signal noise, wiper blade noise [6]), the multiple-model method is more effective than adaptation techniques or online estimation of noise components in terms of computational complexity. In time-varying scenarios, it is also possible to employ *Environmental Sniffing* to detect, track and characterize the noise types [6]. If a clean mixture model is considered as one of the multiple models, the performance of the recognition system can be maintained under high signal-to-noise ratio (SNR) conditions.

19.5 Noise Transition Model

The amount of computation for model-based feature compensation depends primarily on the number of Gaussian components to be computed. The computational expense increases in proportion to the number of multiple models employed for the model interpolation method as described in Section 19.4. However, more accurate modeling for noisy conditions requires a larger number of GMMs with sufficient sized PDFs. Next, we describe a noise transition model to be employed in an effort to reduce the computational complexity.

The motivation is that there might be a smaller sized set of noise types among all types of noise, which we need to consider at a certain time frame or session when employing multiple noise models for PCGMM-based feature compensation. This can reduce the computational expenses. In the *Environmental Sniffing* scheme [6], a *noise language model* was employed to decode the most likely sequence of noise types [6]. Here, the noise transition model is motivated from the noise language model. In order to build the noise language model, in-vehicle acoustic data (i.e., in a Blazer SUV) was collected during a 17-mile route driving which contains samples of all driving conditions expected for use in city and rural areas and then the primary noise conditions were identified as follows:

(1) N1: idle noise, no movement, windows closed
(2) N2: city driving without traffic, windows closed
(3) N3: city driving with traffic, windows closed
(4) N4: highway driving, windows closed
(5) N5: highway driving, windows 2 inches open
(6) N6: highway driving, windows half-way down
(7) N7: windows 2 inches open in city traffic
(8) NX: others

A bigram type of noise language model was constructed using the CMU-Cambridge Statistical Language Modeling (SLM) Toolkit. Here, the connectivity among the noise conditions was employed for the transition model, and we therefore are not considering transition probabilities (e.g., possible

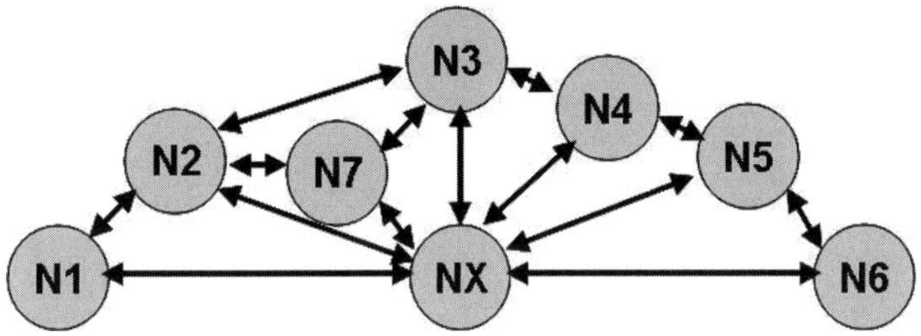

Fig. 19.2 Noise transition model

transitions are either "1" = yes/possible or "0" = no/not possible). Figure 19.2 shows the noise transition model considered in this chapter.

The transition model is applied to the current speech input based on the type of noise observed at the previous utterance when the current speech was produced in a continuous driving condition from the previous utterance. Suppose that the N4 condition (highway driving, windows closed) was determined at the previous utterance by the accumulated posterior probability with this, only four type conditions (N3, N4, N5 and NX) are considered for multiple environmental models by setting the prior probabilities $P(G_i)$ in Eq. (19.6) for the other four conditions (N1, N2, N6 and N7) to zero according to the connectivity of noise transition model as shown in Fig. 19.2. In this case, we expect to have a reduced computational expense compared to the case of not employing a noise transition model (i.e., fully connected noise models).

19.6 Experimental Results

As test data for performance evaluation, the connected single digits portion of the numerous speakers from the CU-Move corpus was selected. This is an identical task to the Aurora2 evaluation framework, so the Aurora2 evaluation toolkit was used to evaluate system performance [10]. The task is connected American English digits consisting of 11 words. Each whole word is represented by a continuous density HMM with 16 states and 3 mixtures per state. In addition to the digits, two silence models (i.e., normal silence and short pause) are used.

The feature extraction algorithm suggested by the European Telecommunication Standards Institute (ETSI) was employed for the experiments [11]. The zeroth cepstral coefficient was used instead of log energy, for the sake of convenience in model combination implementation. After extracting the 13-dimensional MFCC cepstrum, the first-and second-order time derivatives are included during the decoding procedure (a total of 39-dimensional feature vector).

Table 19.1 Performance of baseline system and existing methods on the CU-Move corpus (WER, %)

Baseline	64.77
SS	55.13
SS + CMN	40.39
AFE	31.73
VTS	33.52
VTS + SS + CMN	26.33

The HMM parameters were estimated using 8,840 clean speech training samples included in Aurora2 and performance was evaluated on the selected test set of CU-Move corpus. The test set consists of 464 utterances (length of 50.0 min) spoken by 10 different speakers (5 males and 5 females) in real-life in-vehicle conditions, which were collected in Minneapolis, MN [2]. Data has been down-sampled to 8.0 kHz and reflected a 9.50-dB SNR on average which was obtained by NIST Speech Quality Assurance software [12]. We used eight different types of noise samples (total amount of 2 hours) to train noise models, which were discussed in Section 19.5.

The performance of the baseline system (no compensation) is examined with comparison to several existing preprocessing algorithms in terms of environmental robustness for speech recognition. Spectral subtraction (SS) and cepstral mean normalization (CMN) were selected as conventional algorithms. They represent the most commonly used techniques for additive noise suppression and removal of channel distortion, respectively. In spectral subtraction, the subtraction factor and flooring factor are set at 4.0 and 0.2, respectively, and background noise is estimated using the minimum statistics method with a time delay of approximately 250 ms. For cepstral mean normalization, the average value of the cepstrum over the current input utterance was subtracted from each frame. The AFE (advanced front-end) algorithm developed by ETSI was also evaluated as one of state-of-the-art methods, which contains an iterative Wiener filter and cepstral histogram equalization [13]. We also evaluated another feature compensation method, VTS (vector Taylor series) algorithm for performance comparison where the noisy speech GMM is adaptively estimated using the EM algorithm over each test utterance [9]. Table 19.1 summarizes performance of the baseline system and existing algorithms.

The performance of the PCGMM-based scheme was evaluated using identical conditions to the baseline test. The GMM of the clean speech for PCGMM was estimated using clean speech samples identical to those used for training the HMM. The clean speech model consists of 128 Gaussian components with diagonal covariance matrices. The noise model used for model combination has a single Gaussian model and its prior model was obtained by off-line training. For the prior noise model for single model PCGMM, the noise signals from the type NX were used, which has connections between all other noise

Table 19.2 Performance of PCGMM-based methods

	WER (%)	Relative (%)
PCGMM	62.31	3.80
PCGMMm	33.64	48.06
PCGMMm + SS + CMN	25.38	60.82

Table 19.3 Performance of multi-model PCGMM-based methods

	WER (%)	Relative (%)
IM-PCGMM	34.08	47.38
IM-PCGMM + SS + CMN	25.83	60.12

types. For comparison, we examined the performance in the following combinations:

(1) **PCGMM**: PGCMM-based feature compensation method using model combination of clean speech model and prior noise model trained off-line.
(2) **PCGMMm**: The mean of noise model is updated with the sample mean of silence of each test utterance for PCGMM method. Approximately 200 ms duration of the silence is assumed to exist prior to the beginning of speech in every test utterance.
(3) **PCGMMm + SS + CMN**: PCGMMm method combined with spectral subtraction and CMN

As presented in Table 19.2, the PCGMM-based feature compensation method is effective for in-vehicle conditions, and superior performance of the PCGMM method is demonstrated compared to spectral subtraction combined with CMN in Table 19.1. The results prove that model combination used for estimation of the noisy speech GMM is effective in representing the noise corruption process. Relative improvement of 48.06% over the baseline in WER was obtained through updating the mean of the noise model (PCGMMm), which is better or comparable to AFE and single VTS. The PCGMMm method combined with spectral subtraction and CMN has a relative improvement of 60.82% in WER, and outperforms all other existing methods.

Using the same setup, performance evaluation of the multi-model schemes for PCGMM was also conducted. In the interpolation of multi-model PCGMM method, nine types of noise models (N1, N2, ... , N7 and NX including clean condition) were used for model combination to generate noisy speech GMMs. As presented in Table 19.3, we see that PCGMM-based feature compensation schemes with the interpolation method of multiple models are effective for in-vehicle conditions, with superior performance over existing conventional algorithms. The PCGMM-based feature compensation with interpolated models (IM-PCGMM) presents comparable performance to the

Table 19.4 Performance of multi-model PCGMM-based methods with noise transition model

	WER (%)	Relative (%)
IM-PCGMM	33.86	47.72
IM-PCGMM + SS + CMN	25.98	59.89

Table 19.5. Computational reduction of multi-model PCGMM-based methods by employing noise transition model

	No. of activated noise models	Computational reduction (%)
IM-PCGMM	6.22	30.89
IM-PCGMM + SS + CMN	6.59	26.78

single adapted model approach (PCGMMm) which is shown in Table 19.2. This proves that interpolation of multiple models is very effective for compensating the feature adaptively under blind noisy environments and changing noise types in every utterance. A significant improvement was obtained by combining the IM-PCGMM method with spectral subtraction and CMN.

Tables 19.4 and 19.5 present the performance of the IM-PCGMM-based method employing the noise transition model described in Section 19.5. The test utterances are submitted to the speech recognizer in the same time-order as recorded in-vehicle. With the noise transition model, for a particular speaker, it is determined which noise types are considered for the current utterance for multiple environmental models, based on the noise condition which has the highest score (*a posteriori*) at the previous utterance. From Table 19.4, IM-PCGMM with the noise transition model demonstrates comparable performance versus the case of a fully connected noise model in Table 19.3. In order to investigate the relationship between performance and computational expense brought by employing the noise transition model, the average number of activated noise models and resulting computational reduction are presented in Table 19.5.

The computational reduction was calculated compared with a fully connected noise model which has nine activated noise conditions. From the results, it was found that employing the noise transition model is useful for reducing the computational complexity (e.g., up to 30.89% savings in computation) while holding the original performance at comparable levels.

19.7 Conclusion

In this chapter, we evaluated the PCGMM-based feature compensation method on the CU-Move in-vehicle speech corpus which contains a range of background noise observed in real-life in-vehicle conditions. To reduce the computational complexity, a noise transition model was proposed for a multiple model speech recognition approach. Experimental results demonstrate that our feature compensation method is effective in accomplishing reliable and efficient speech recognition in actual in-vehicle environments.

References

1. http://www.utdallas.edu/research/utdrive.
2. J.H.L. Hansen, X. Zhang, M. Akbacak, U. Yapanel, B. Pellom, W. Ward, and P. Angkititrakul, "CU-Move: Advanced In-Vehicle Speech Systems for Route Navigation," *DSP for In-Vehicle and Mobile Systems*, Springer, 2004.
3. X. Zhang and J.H.L. Hansen, "CSA-BF: A Constrained Switched Adaptive Beamformer for Speech Enhancement and Recognition in Real Car Environments," *IEEE Transactions on Speech and Audio Processing*, vol. 11, no. 6, pp. 733–745, 2003.
4. W. Kim, S. Ahn, and H. Ko, "Feature Compensation Scheme Based on Parallel Combined Mixture Model," *Eurospeech2003*, pp. 667–680, 2003.
5. W. Kim, O. Kwon, and H. Ko, "PCMM-Based Feature Compensation Scheme Using Model Interpolation and Mixture Sharing," *ICASSP2004*, pp. 989–992, 2004.
6. M. Akbacak and J.H.L. Hansen, "Environmental Sniffing: Noise Knowledge Estimation for Robust Speech Systems," *IEEE Transactions on Audio, Speech and Language Processing*, vol.15, no 2, pp. 465–477, 2007.
7. A.P. Varga and R.K. Moore, "Hidden Markov Model Decomposition of Speech and Noise," *ICASSP90*, pp. 845–848, 1990.
8. M.J.F. Gales and S.J. Young, "Robust Continuous Speech Recognition Using Parallel Model Combination," *IEEE Transactions on Speech and Audio Processing*, vol. 4, no. 5, pp. 352–359, 1996.
9. P.J. Moreno, B. Raj, and R.M. Stern, "Data-Driven Environmental Compensation for Speech Recognition: A Unified Approach," *Speech Communication*, vol. 24, pp. 267–85, 1998.
10. H.G. Hirsch and D. Pearce, "The AURORA Experimental Framework for the Performance Evaluations of Speech Recognition Systems Under Noisy Conditions," *ISCA ITRW ASR2000*, Sept. 2000.
11. ETSI Standard Doc., "Speech Processing, Transmission and Quality Aspects (STQ); Distributed Speech Recognition; Front-End Feature Extraction Algorithm; Compression Algorithms," *ETSI ES 201 108 v1.1.2 (2000–04)*, 2000.
12. NIST SPeech Quality Assurance (SPQA) package version 2.3, http://www.nist.gov/speech.
13. ETSI Standard Doc., "Speech Processing, Transmission and Quality Aspects (STQ); Distributed Speech Recognition; Advanced Front-End Feature Extraction Algorithm; Compression Algorithms," *ETSI ES 202 050 v1.1.1 (2002–10)*, 2002.

Index

A
Abnormal driving, 24
Acoustic and throat microphone recordings, 162, 163–166
Active shape model, 212
Adaptive cruise control (ACC), 11, 12–13, 14, 45–46
Additive white Gaussian noise (AWGN), 140, 141
Ad hoc wireless networks, 83, 84, 85, 87, 88
Array processing, 140, 141, 142, 188, 234
Attention, 3–5, 32, 56, 150, 154–155, 172
Audio–visual speech interface (AVSI), 212, 214, 215, 216, 218
Auditory displays, 4, 6
Automatic repeat request, 86, 87–88
Automatic speech recognition (ASR), 28, 29, 30, 57, 58, 61, 172, 211–212, 217, 218, 221–222, 230–231, 237
Autonomous navigation, 2–3, 137
Autoregressive adaptation, 221, 222, 226, 227, 229, 230, 231

B
Bayesian discrimination, 112, 114–117
Beamformer, 137, 143–147, 192, 193, 194
Brake pedal pressure sensor, 25, 26, 27–28, 30, 59
BSSID, 70, 78

C
CAN-Bus, 24, 25, 27, 30, 31, 57, 59, 60–61, 62–63, 66
Category estimation, 113, 116–117, 123
Coherence, 153, 154, 156, 157
Collision mitigation, 11, 12, 13
Complex background, 112, 123
  clothing style, 123
Conferencing applications, 84, 87
Confusion matrix, 131, 132, 133, 204, 205, 206, 207, 208, 209, 210
Cooperating vehicles, 16–20
Coupling, 107–108, 109
Crash avoidance technology, 1–2, 5
Cross-layer communications, 87, 88
Cross-probability model, 171–184
CU-Move, 234, 235, 239, 240

D
Data collection, 23–43, 45–53, 57, 58, 60–61, 126–128, 199
  vehicle, 25, 46–51, 58, 61, 199
Dedicated short-range communication (DSRC), 14, 19
3D head tracking, 37–39
Dialogue rhythm, 197–198
Distraction, 4, 5–6, 32, 42, 56, 57, 61, 63–65, 153, 172, 198
Driver-adaptive systems, 10, 13–16
Driver behavior, 6, 24, 38, 40–41, 56–57, 58, 62, 65, 89
Driver modeling, 24, 42, 65
Drive-Safe Consortium, 24, 42
Driving assistance, 9–21, 13, 24, 83–84, 111–112
Driving signal, 197–210
Drowsiness, 13, 14, 38, 97–109

E
EEG, 25, 26, 27, 32, 34, 98, 109, 125–147
Elderly drivers, 10, 15, 16, 17, 18, 20, 21
Emotion recognition, 125–147
Error propagation, 228
Expressway, 46, 53
Eye movement, 65, 150, 151–154, 155, 156, 158

F
Facial action unit, 101
Facial expressions, 36, 98, 100–102, 108–109
Facial feature tracking, 25, 36–37, 38
Fatigue, driver, 32, 36, 38, 42, 98, 99, 109
Feature vector normalization, 171–184
combination, 187–196
compensation, 233–243
Fresnel's approximation, 139
Fuzzy neural network, 125, 126, 130–131

G
Gas pedal pressure sensor, 27, 59
GMM (Gaussian mixture model), 42, 56, 171–184, 190, 193–194, 195, 234, 235, 236, 237, 240, 241
GPS (global positioning system), 19, 25, 27–28, 32, 57, 60, 61, 69, 71, 72–73, 76, 80, 199
GSM (global system for mobile communications), 69, 70, 80

H
Hand-held wireless device, 6
Head movements, 39, 58, 98, 100, 101
Hidden Markov model (HMM), 34, 35, 36, 165–166, 188, 212, 230–231
Highway safety, 1, 2
HMM-based correlation model, 165
pattern recognition, 198, 200–201
Human factors, 2, 3, 5, 56
Huygens' Principle, 140
Hybrid integration, 216
Hypothesis, 188, 190–191, 194–195, 224, 226

I
Infrared image, 12, 14, 112, 113
Intelligent transport systems (ITS), 1–2, 9, 13, 14–15
Intervehicle
communications, 15, 19, 84
transmissions, 89
In-vehicle information systems, 2, 5, 55–56, 172, 233–243
condition, 234, 238, 240, 241–242
Istanbul route, 28
IVMCTool, 23

K
Kernel density estimator, 112, 113, 114–117, 118

L
Lane keeping, 11, 13, 45–46, 56–57, 65
Laser ranger finder, 25, 26, 27, 28, 31, 34, 39
Lidar, 11, 12–13, 14, 138
Lip-reading, 212
Localization, 18–19, 21, 69–73, 76–80
Location information processing, 69–81
Locky.jp, 70, 76, 78–79
Low-delay applications, 87

M
Machine learning, 97–109
Mel-frequency cepstrum coefficients (MFCCs), 35, 42, 175–176, 181, 188–190, 191, 192–194, 212, 215, 239
MEMLIN, 172–175, 176–177, 179, 180, 181–182, 183–184
Mental work load, 149–158
Metropolitan-scale location services, 70, 76–80
MFCC, *see* Mel-frequency cepstrum coefficients (MFCCs)
Microphone array, 57–58, 187–196
Minimum mean square error estimation (MMSE), 172–173, 175, 222, 223, 224, 225–226, 236
MLP (multilayer perceptron), 130, 131, 132, 133, 134, 135
Model-based approach, 86–87
Model combination, 216–217, 233–243
Mouth model, 213
Multi-classifier, 33–36
Multimedia communications, 85–86, 87–88, 89–90
Multimedia database, 46–53
Multimodal driving corpus, 46, 57–61
Multinomial logistic regression (MLR), 98, 105, 106, 107
Multistream features, 215, 216

N
Navigational dialog, 28, 29
Navigation systems, 10, 11–12, 13, 14–15, 45–46, 172, 192, 197–198, 199, 211–212
Near-field model, 139
NEDO, 24, 25, 28, 57
Noise spectrum estimation, 221–231
transition model, 234, 238–239
NTSB (National Transportation Safety Board), 1–2

O
Obstacle detection, 12, 84, 138, 142
On-line banking, 28, 29, 30
Orientation estimation, 70, 73–76

P
Packet scheduling, 83–95
Pedestrian extraction, 113
Perception, 3–5, 10, 17, 135
Perceptually optimized scheduling, 83–95
Periodogram, 224, 228
Physiological signal, 46, 50, 98
Place Engine, 79–80
PlaceLab, 79–80
Posture estimation, 112, 113, 118, 122
Probability density function (pdf), 112, 113, 172, 175, 176, 177, 178, 223, 225, 227, 238
Probability map template, 112, 118–123
Proximity, 12, 70, 71

Q
Quantitative evaluation, 149–158

R
Reaction time, 4, 156
Real-world driving, 53
Reference driving, 29
Retransmission mechanism, 87–88, 90
Road-sign reading, 28, 29
Robust speech processing, 161–162
  automatic speech recognition, 172
Route navigation dialog, 55–56, 234
RVM (relevance vector machine), 126, 129–130, 131, 133, 134

S
Safety, 1–7, 9, 13, 15, 17, 38, 45–46, 56, 60, 61, 63, 84, 98, 171–172, 197–198
  signals, 84
Scene analysis, 70–71, 72
Semicircular canals, 151–152
Shape-map, 112, 118, 119–120, 121–122
Short-time Fourier transform (STFT), 222, 223, 224, 225
Signal strength, 70, 71, 73–74, 75–76, 78
Speech recognition
  enhancement, 161–162, 222
  interface, 2–3, 211
Spoken dialog, 46, 52, 65, 197–198
Statoliths, 151–152
Steering assistance, 11
Support vector machine (SVM), 34–35, 101–102, 129–130, 131, 135
Synchronous recording, 46, 50–51

T
Template matching, 118–123
Texture, 138, 139, 143–146
Tied-mixture hidden Markov model (TMHMM), 212, 216, 217
Time-of-flight (TOF), 139, 141

U
Ultrasound imaging, 137–147
Urban road, 53, 192
UTDrive, 55–66
UYANIK, 23–43

V
Variance normalization, 188, 189–190, 192, 193, 194, 195
Verbal *n*-back task, 151–152, 155
Vestibulo-ocular reflex (VOR), 150, 151–152, 153, 155, 156
Voice activity detection (VAD), 222, 228–229, 230

W
WAVE, 84, 87
Wireless LAN, 69–81
Wireless location systems, 69–81
Word error rate (WER), 177, 181, 217, 218, 222, 229, 230–231, 240, 241, 242
Working memory, 154–156, 157

Printed in the United States of America